SpringerBriefs in Mathematics

SpringerBriefs in Mathematics showcases expositions in all areas of mathematics and applied mathematics. Manuscripts presenting new results or a single new result in a classical field, new field, or an emerging topic, applications, or bridges between new results and already published works, are encouraged. The series is intended for mathematicians and applied mathematicians.

BCAM SpringerBriefs

BCAM *SpringerBriefs* aims to publish contributions in the following disciplines: Applied Mathematics, Finance, Statistics and Computer Science. BCAM has appointed an Editorial Board, who evaluate and review proposals.

Typical topics include: a timely report of state-of-the-art analytical techniques, bridge between new research results published in journal articles and a contextual literature review, a snapshot of a hot or emerging topic, a presentation of core concepts that students must understand in order to make independent contributions.

Please submit your proposal to the Editorial Board or to Francesca Bonadei, Executive Editor Mathematics, Statistics, and Engineering: francesca.bonadei@springer.com.

basque center for applied **mathematics**

More information about this series at http://www.springer.com/series/10030

Jingrui Sun · Jiongmin Yong

Stochastic Linear-Quadratic Optimal Control Theory: Open-Loop and Closed-Loop Solutions

Jingrui Sun
Department of Mathematics
Southern University of Science
and Technology
Shenzhen, Guangdong, China

Jiongmin Yong
Department of Mathematics
University of Central Florida
Orlando, FL, USA

ISSN 2191-8198   ISSN 2191-8201 (electronic)
SpringerBriefs in Mathematics
ISBN 978-3-030-20921-6   ISBN 978-3-030-20922-3 (eBook)
https://doi.org/10.1007/978-3-030-20922-3

This Springer imprint is published by the registered company Springer Nature Switzerland AG
The registered company address is: Gewerbestrasse 11, 6330 Cham, Switzerland

To Our Parents

Yuqi Sun and Xiuying Ma

Wenyao Yong and Xiangxia Chen

Preface

Linear-quadratic optimal control theory (LQ theory, for short) has a long history, and the general consensus is that LQ theory is quite mature. It chiefly involves three well-known and relevant issues: the existence of optimal controls, the solvability of the optimality system (which is a two-point boundary value problem), and the solvability of the associated Riccati equation. Broadly speaking, these three issues are somehow equivalent. For the past few years we, together with our collaborators, have been reinvestigating LQ theory for stochastic systems with deterministic coefficients. In this context, we have identified a number of interesting issues, including

- For finite-horizon LQ problems, open-loop optimal controls may not have a closed-loop representation.
- For finite-horizon LQ problems, a distinction should be made between open-loop optimal controls and closed-loop optimal strategies. The existence of the latter implies the existence of the former, but not vice versa.
- For infinite-horizon LQ problems (with constant coefficients), under proper conditions, the open-loop and the closed-loop solvability are equivalent.

Moreover, our investigations have revealed some previously unknown aspects; these include but are not limited to the following:

- For finite-horizon LQ problems, the open-loop solvability is equivalent to the solvability of the optimality system, which is a forward–backward stochastic differential equation (FBSDE), together with the convexity of the cost functional.
- For finite-horizon LQ problems, the closed-loop solvability is equivalent to the existence of a regular solution to the Riccati differential equation.
- For infinite-horizon LQ problems (with constant coefficients), both the open-loop and the closed-loop solvability are equivalent to the solvability of an algebraic Riccati equation.

The purpose of this book is to systematically present the above-mentioned results and many other relevant ones. We assume that readers are familiar with basic stochastic analysis and stochastic control theory.

This work was supported in part by NSFC Grant 11901280 and NSF Grants DMS-1406776 and DMS-1812921.

The authors would also like to express their gratitude to the anonymous referees for their constructive comments, which led to this improved version.

Shenzhen, China   Jingrui Sun
Orlando, USA   Jiongmin Yong
March 2020

Contents

About the Authors

Jingrui Sun received his Ph.D. in Mathematics from the University of Science and Technology of China in 2015. From 2015 to 2017, he was a Postdoctoral Fellow at the Hong Kong Polytechnic University and then a Research Fellow at the National University of Singapore. From 2017 to 2018, he was a Visiting Assistant Professor at the University of Central Florida, USA. Since the spring of 2019, he has been an Assistant Professor at the Southern University of Science and Technology, China. Dr. Sun has broad interests in the area of control theory and its applications. Aside from his primary research on stochastic optimal control and differential games, he is exploring forward and backward stochastic differential equations, stochastic analysis, and mathematical finance.

Jiongmin Yong received his Ph.D. from Purdue University in 1986 and is currently a Professor of Mathematics at the University of Central Florida, USA. His main research interests include stochastic control, stochastic differential equations, and optimal control of partial differential equations. Professor Yong has co-authored the following influential books: "Stochastic Control: Hamiltonian Systems and HJB Equations" (with X. Y. Zhou, Springer 1999), "Forward-Backward Stochastic Differential Equations and Their Applications" (with J. Ma, Springer 1999), and "Optimal Control Theory for Infinite-Dimensional Systems" (with X. Li, Birkhauser 1995). His current interests include time-inconsistent stochastic control problems.

Frequently Used Notation

I. Notation for Euclidean Spaces and Matrices

(1) $\mathbb{R}^{n \times m}$: the space of all $n \times m$ real matrices.

(2) $\mathbb{R}^n = \mathbb{R}^{n \times 1}$; $\mathbb{R} = \mathbb{R}^1$; $\overline{\mathbb{R}} = [-\infty, \infty]$.

(3) $\mathbb{S}^n$: the space of all symmetric $n \times n$ real matrices.

(4) $\mathbb{S}_+^n$: the subset of $\mathbb{S}^n$ consisting of positive definite matrices.

(5) $\overline{\mathbb{S}}_+^n$: the subset of $\mathbb{S}^n$ consisting of positive semi-definite matrices.

(6) I_n: the identity matrix of size n, which is also denoted simply by I if no confusion occurs.

(7) $M^\top$: the transpose of a matrix M.

(8) $M^\dagger$: the Moore-Penrose pseudoinverse of a matrix M.

(9) $\mathrm{tr}(M)$: the sum of diagonal elements of a square matrix M, called the trace of M.

(10) $\langle \cdot, \cdot \rangle$: the inner product on a Hilbert space. In particular, the usual inner product on $\mathbb{R}^{n \times m}$ is given by $\langle M, N \rangle \longmapsto \mathrm{tr}(M^\top N)$.

(11) $|M| \triangleq \sqrt{\mathrm{tr}(M^\top M)}$: the Frobenius norm of a matrix M.

(12) $\mathscr{R}(M)$: the range of a matrix or an operator M.

(13) $\mathscr{N}(M)$: the kernel of a matrix or an operator M.

(14) $A \geqslant B$: $A - B$ is a positive semi-definite symmetric matrix.

(15) $\mathcal{Q}(P) \triangleq PA + A^\top P + C^\top PC + Q$; See Sects. 2.4 and 3.3.1.

(16) $\mathcal{S}(P) \triangleq B^\top P + D^\top PC + S$; See Sects. 2.4 and 3.3.1.

(17) $\mathcal{R}(P) \triangleq R + D^\top PD$; See Sects. 2.4 and 3.3.1.

II. Sets and Spaces of Functions and Processes

Let $\mathbb{H}$ be a Euclidian space (which could be $\mathbb{R}^n$, $\mathbb{R}^{n \times m}$, etc.).

(1) $C([t, T]; \mathbb{H})$: the space of $\mathbb{H}$-valued, continuous functions on $[t, T]$.

(2) $L^p(t, T; \mathbb{H})$: the space of $\mathbb{H}$-valued functions that are pth $(1 \leqslant p < \infty)$ power Lebesgue integrable on $[t, T]$.

(3) $L^\infty(t, T; \mathbb{H})$: the space of $\mathbb{H}$-valued, Lebesgue measurable functions that are essentially bounded on $[t, T]$.

(4) $L^2_{\mathcal{F}_t}(\Omega; \mathbb{H})$: the space of $\mathcal{F}_t$-measurable, $\mathbb{H}$-valued random variables ξ such that $\mathbb{E}|\xi|^2 < \infty$.

(5) $L^2_{\mathbb{F}}(\Omega; L^1(t, T; \mathbb{H}))$: the space of $\mathbb{F}$-progressively measurable, $\mathbb{H}$-valued processes $\varphi : [t, T] \times \Omega \to \mathbb{H}$ such that $\mathbb{E}\left[\int_t^T |\varphi(s)| ds \right]^2 < \infty$.

(6) $L^2_{\mathbb{F}}(t, T; \mathbb{H})$: the space of $\mathbb{F}$-progressively measurable, $\mathbb{H}$-valued processes $\varphi : [t, T] \times \Omega \to \mathbb{H}$ such that $\mathbb{E} \int_t^T |\varphi(s)|^2 ds < \infty$.

(7) $L^2_{\mathbb{F}}(\mathbb{H})$: the space of $\mathbb{F}$-progressively measurable, $\mathbb{H}$-valued processes $\varphi : [0, \infty) \times \Omega \to \mathbb{H}$ such that $\mathbb{E} \int_0^\infty |\varphi(t)|^2 dt < \infty$.

(8) $L^2_{\mathbb{F}}(\Omega; C([t, T]; \mathbb{H}))$: the space of $\mathbb{F}$-adapted, continuous, $\mathbb{H}$-valued processes $\varphi : [t, T] \times \Omega \to \mathbb{H}$ such that $\mathbb{E}\left[\sup_{s \in [t,T]} |\varphi(s)|^2 \right] < \infty$.

(9) $\mathcal{X}_t = L^2_{\mathcal{F}_t}(\Omega; \mathbb{R}^n)$.

(10) $\mathcal{X}[t, T] = L^2_{\mathbb{F}}(\Omega; C([t, T]; \mathbb{R}^n))$.

(11) $\mathcal{U}[t, T] = L^2_{\mathbb{F}}(t, T; \mathbb{R}^m)$.

(12) $\mathcal{X}_{loc}[0, \infty) = \bigcap_{T > 0} \mathcal{X}[0, T]$.

(13) $\mathcal{X}[0, \infty)$: the subspace of $\mathcal{X}_{loc}[0, \infty)$ consisting of processes φ which are square-integrable: $\mathbb{E} \int_0^\infty |\varphi(t)|^2 dt < \infty$.

Chapter 1
Introduction

Abstract This chapter is an introduction to the linear-quadratic optimal control problem and serves as a motivation for the book. The history of linear-quadratic problems is briey reviewed and some classical results for deterministic linear-quadratic problems are collected. As an abstract framework for studying the linear-quadratic problem, the optimization problem for quadratic functionals in a Hilbert space is also discussed.

Keywords Linear-quadratic · Optimal control · Value function · Optimality system · Riccati equation · Quadratic functional

1.1 Why Linear-Quadratic Problems?

Let $(\Omega, \mathcal{F}, \mathbb{F}, \mathbb{P})$ be a complete filtered probability space on which a one-dimensional standard Brownian motion $W = \{W(t); 0 \leqslant t < \infty\}$ is defined such that $\mathbb{F} \equiv \{\mathcal{F}_t\}_{t \geqslant 0}$ is its natural filtration augmented by all the $\mathbb{P}$-null sets in $\mathcal{F}$. Consider the following controlled stochastic differential equation (SDE, for short):

$$\begin{cases} dX(s) = b(s, X(s), u(s))ds + \sigma(s, X(s), u(s))dW(s), \quad s \in [t, T], \\ X(t) = x. \end{cases} \tag{1.1.1}$$

In the above, X is called the *state process*, u is called the *control process*, b and σ are given maps. Under some mild conditions, for any *initial pair* $(t, x) \in [0, T) \times \mathbb{R}^n$ and control process u selected from some set $\mathcal{U}[t, T]$, the above *state equation* admits a unique solution $X(\cdot) \equiv X(\cdot; t, x, u)$. To measure the performance of the control u, we introduce the following *cost functional*:

$$J(t, x; u) = \mathbb{E}\left[\int_t^T f(s, X(s), u(s))ds + g(X(T)) \right]. \tag{1.1.2}$$

A classical stochastic optimal control problem is to minimize the cost functional (1.1.2) subject to the state equation (1.1.1) by selecting u from $\mathcal{U}[t, T]$.

Suppose all the involved functions are differentiable, and after possible change of coordinates, we assume that the state process X lies near the equilibrium $x = 0$, and for the control process, $u = 0$ represents "no action". By Taylor expansion, we have

$$b(s, x, u) = b(s, 0, 0) + b_x(s, 0, 0)x + b_u(s, 0, 0)u + \cdots,$$

$$\sigma(s, x, u) = \sigma(s, 0, 0) + \sigma_x(s, 0, 0)x + \sigma_u(s, 0, 0)u + \cdots,$$

$$f(s, x, u) = f(s, 0, 0) + f_x(s, 0, 0)x + f_u(s, 0, 0)u + \frac{1}{2}\langle f_{xx}(s, 0, 0)x, x\rangle$$

$$+ \langle f_{xu}(s, 0, 0)x, u\rangle + \frac{1}{2}\langle f_{uu}(s, 0, 0)u, u\rangle + \cdots,$$

$$g(x) = g(0) + g_x(0)x + \frac{1}{2}\langle g_{xx}(0)x, x\rangle + \cdots.$$

In the above, "$\cdots$" stands for some higher order terms. If we neglect those higher order terms and denote

$$
\begin{array}{lll}
A(s) = b_x(s, 0, 0), & B(s) = b_u(s, 0, 0), & b(s) = b(s, 0, 0), \\
C(s) = \sigma_x(s, 0, 0), & D(s) = \sigma_u(s, 0, 0), & \sigma(s) = \sigma(s, 0, 0), \\
Q(s) = f_{xx}(s, 0, 0), & S(s) = f_{xu}(s, 0, 0), & R(s) = f_{uu}(s, 0, 0), \\
q(s) = f_x(s, 0, 0)^\top, & \rho(s) = f_u(s, 0, 0)^\top, & G = g_{xx}(0), \; g = g_x(0)^\top,
\end{array}
$$

then the state equation (1.1.1) becomes

$$
\begin{cases}
dX(s) = [A(s)X(s) + B(s)u(s) + b(s)]ds \\
\qquad\qquad + [C(s)X(s) + D(s)u(s) + \sigma(s)]dW(s), \quad s \in [t, T], \qquad (1.1.3) \\
X(t) = x,
\end{cases}
$$

and the cost functional (1.1.2) becomes

$$J(t, x; u) = \frac{1}{2}\mathbb{E}\left\{ \int_t^T \left[\langle Q(s)X(s), X(s)\rangle + 2\langle S(s)X(s), u(s)\rangle \right.\right.$$

$$\left. + \langle R(s)u(s), u(s)\rangle + 2\langle q(s), X(s)\rangle + 2\langle \rho(s), u(s)\rangle \right]ds$$

$$\left. + \langle GX(T), X(T)\rangle + 2\langle g, X(T)\rangle + 2\int_t^T f(s, 0, 0)ds + 2g(0)\right\}.$$

Hence, the original stochastic optimal control problem is approximately equivalent to minimizing

$$J(t, x; u) = \mathbb{E}\bigg\{ \int_t^T \Big[\langle Q(s)X(s), X(s) \rangle + 2\langle S(s)X(s), u(s) \rangle$$
$$+ \langle R(s)u(s), u(s) \rangle + 2\langle q(s), X(s) \rangle + 2\langle \rho(s), u(s) \rangle \Big] ds$$
$$+ \langle GX(T), X(T) \rangle + 2\langle g, X(T) \rangle \bigg\},$$

which is up to the quadratic terms in (X, u), subject to the linear state equation (1.1.3). We refer to such a problem as a *linear-quadratic* stochastic optimal control problem (LQ problem, for short). We refer to A, B, C, and D as the *coefficients*, b and σ in (1.1.3) as the *nonhomogeneous terms*, and (t, x) as the *initial pair* of the state equation; and refer to Q, S, R, q, ρ, G, g as the *weights* of the cost functional.

We have seen that the above stochastic LQ problem is an approximation of general nonlinear stochastic optimal control problems. There are two major advantages of the LQ problem: Mathematically, the problem with such a special structure will lead to some much nicer results than the general ones; and in real applications, such an approximation is well-acceptable and good enough for many concrete problems.

The study of LQ problems can be traced back to the works of Bellman–Glicksberg–Gross [6] in 1958, Kalman [24] in 1960 and Letov [27] in 1961. The above-mentioned works were concerned with deterministic cases, i.e., the state equation is a linear ordinary differential equation (ODE, for short), and all the involved functions are deterministic. For such a case, it is known that $R(s) \geqslant 0$, meaning that $R(s)$ is positive semi-definite for almost every $s \in [0, T]$, is necessary for the corresponding LQ problem to be finite (meaning that the infimum of the cost functional is finite). When the control weighting matrix $R(s) \geqslant \delta I$ for some $\delta > 0$, meaning that $R(s)$ is uniformly positive definite for almost every $s \in [0, T]$, then, under some mild additional conditions on the other weighting coefficients, the problem can be solved elegantly via the Riccati equation; see Anderson–Moore [4] for a thorough presentation of the Riccati equation approach (see also Yong–Zhou [62]). Stochastic LQ problems were firstly studied by Wonham [52] in 1968, followed by several researchers (see, for example, Davis [17] and Bensoussan [7]). In those works, the assumption $R(s) \geqslant \delta I$ ($\delta > 0$) was still taken for granted. More precisely, under the *standard condition* that

$$G \geqslant 0, \quad R(s) \geqslant \delta I, \quad Q(s) - S(s)^\top R(s)^{-1} S(s) \geqslant 0, \quad \text{a.e. } s \in [0, T],$$

for some $\delta > 0$, the corresponding Riccati equation is uniquely solvable and the corresponding stochastic LQ problem admits a unique optimal control which has a linear state feedback representation (see [62, Chap. 6]).

In 1998, Chen–Li–Zhou [11] found that the stochastic LQ problem might still be solvable even if $R(s)$ is not positive semi-definite. See also some follow-up works of Lim–Zhou [31], Chen–Zhou [14], and Chen–Yong [13], as well as the works of McAsey–Mou [35] and Qian–Zhou [42] on the study of solvability of indefinite Riccati equations (under certain technical conditions). In 2001, Ait Rami–Moore–

Zhou [1] introduced a generalized Riccati equation involving the pseudoinverse of a matrix and an additional algebraic constraint; see also Ait Rami–Zhou [2] for stochastic LQ optimal control problems on $[0, \infty)$ and a follow-up work of Wu–Zhou [53]. Since 2010s, Yong [59], Huang–Li–Yong [23], Li–Sun–Yong [30], Sun [43], Wei–Yong–Yu [51], and Li–Sun–Xiong [29] studied LQ problems involving mean-fields. For stochastic LQ optimal control problems with random coefficients, we refer to the works of Chen–Yong [12], Kohlmann–Tang [26], Tang [48, 49], and Sun–Xiong–Yong [46].

1.2   Standard Results for Deterministic LQ Problems

In this section, we briefly recall the results for deterministic LQ problems. This leads to the main motivation of presenting the results in later chapters.

Consider the controlled linear ODE

$$\begin{cases} \dot{X}(s) = AX(s) + Bu(s), \quad s \in [t, T], \\ X(t) = x \end{cases}$$

and the cost functional

$$J_D(t, x; u) = \langle GX(T), X(T) \rangle + \int_t^T \left\langle \begin{pmatrix} Q & S^\top \\ S & R \end{pmatrix} \begin{pmatrix} X(s) \\ u(s) \end{pmatrix}, \begin{pmatrix} X(s) \\ u(s) \end{pmatrix} \right\rangle ds.$$

For simplicity, we only look at the case of constant coefficients, without the nonhomogeneous terms in the state equation and the linear terms in the cost functional. We assume that

$$A \in \mathbb{R}^{n \times n}, \quad B \in \mathbb{R}^{n \times m}, \quad G, Q \in \mathbb{S}^n, \quad S \in \mathbb{R}^{m \times n}, \quad R \in \mathbb{S}^m, \tag{1.2.1}$$

where $\mathbb{R}^{n \times m}$ is the set of $n \times m$ real matrices and $\mathbb{S}^n$ is the set of $n \times n$ symmetric (real) matrices. Let

$$\mathcal{U}_D[t, T] = L^2(t, T; \mathbb{R}^m) \equiv \left\{ u : [t, T] \to \mathbb{R}^m \ \middle| \ \int_t^T |u(s)|^2 ds < \infty \right\}.$$

We now pose the following problem.

Problem (DLQ). For given initial pair $(t, x) \in [0, T) \times \mathbb{R}^n$, find a $\bar{u} \in \mathcal{U}_D[t, T]$ such that

$$J_D(t, x; \bar{u}) = \inf_{u \in \mathcal{U}_D[t,T]} J_D(t, x; u) \equiv V(t, x). \tag{1.2.2}$$

Any $\bar{u} \in \mathcal{U}_D[t, T]$ satisfying (1.2.2) is called an (open-loop) *optimal control* of Problem (DLQ) (for the initial pair (t, x)), the corresponding *state trajectory* $\bar{X}$ is called an (open-loop) *optimal state trajectory*, $(\bar{X}, \bar{u})$ is called an (open-loop) *optimal pair*, and the function V is called the *value function* of Problem (DLQ).

The following proposition summarizes the main results for Problem (DLQ). See Yong–Zhou [62, Chap. 6, Sect. 2] and Yong [60, Chap. 6, Sect. 1]. The result will be stated for constant coefficient case; the time-varying case is essentially the same.

Proposition 1.2.1 *Let* (1.2.1) *hold.*

(i) *If* $V(t, x) > -\infty$ *for some* $(t, x) \in [0, T) \times \mathbb{R}^n$, *then* $R \geqslant 0$.

(ii) *Given* $(t, x) \in [0, T) \times \mathbb{R}^n$. *If Problem* (DLQ) *admits an optimal pair* $(\bar{X}, \bar{u})$, *then the following optimality system is solvable:*

$$
\begin{cases}
\dot{\bar{X}}(s) = A\bar{X}(s) + B\bar{u}(s), \\
\dot{\bar{Y}}(s) = -A^\top \bar{Y}(s) - Q\bar{X}(s) - S^\top \bar{u}(s), \\
\bar{X}(t) = x, \quad \bar{Y}(T) = G\bar{X}(T), \\
B^\top \bar{Y}(s) + S\bar{X}(s) + R\bar{u}(s) = 0,
\end{cases} \tag{1.2.3}
$$

which is a coupled two-point boundary value problem with the coupling resulting from the last equality in the system, called the stationarity condition.

(iii) *Let* $R > 0$ *and suppose that the Riccati equation*

$$
\begin{cases}
\dot{P}(s) + P(s)A + A^\top P(s) + Q \\
\quad - [P(s)B + S^\top]R^{-1}[B^\top P(s) + S] = 0, \quad s \in [0, T], \\
P(T) = G,
\end{cases} \tag{1.2.4}
$$

admit a solution $P : [0, T] \to \mathbb{S}^n$. *For any initial pair* (t, x), *define the control* $\bar{u}$ *by*

$$
\bar{u}(s) = -R^{-1}[B^\top P(s) + S]\bar{X}(s), \quad s \in [t, T],
$$

where $\bar{X}$ *is the solution to the closed-loop system*

$$
\begin{cases}
\dot{\bar{X}}(s) = \{A - BR^{-1}[B^\top P(s) + S]\}\bar{X}(s), \quad s \in [t, T], \\
\bar{X}(t) = x.
\end{cases}
$$

Then $\bar{u}$ *is the unique optimal control, and*

$$
V(t, x) = \langle P(t)x, x \rangle, \quad \forall (t, x) \in [0, T] \times \mathbb{R}^n.
$$

(iv) *If* $R > 0$ *and for any initial pair* $(t, x) \in [0, T) \times \mathbb{R}^n$, *the forward-backward ODE (FBODE, for short)*

$$\begin{cases} \dot{\bar{X}}(s) = (A - BR^{-1}S)\bar{X}(s) - BR^{-1}B^\top \bar{Y}(s), \\ \dot{\bar{Y}}(s) = -(Q - S^\top R^{-1}S)\bar{X}(s) - (A - BR^{-1}S)^\top \bar{Y}(s), \\ \bar{X}(t) = x, \quad \bar{Y}(T) = G\bar{X}(T) \end{cases} \quad (1.2.5)$$

has a unique solution $(\bar{X}, \bar{Y})$, then the Riccati equation (1.2.4) admits a unique solution P. In addition, if the standard condition

$$R > 0, \quad Q - SR^{-1}S^\top \geqslant 0, \quad G \geqslant 0, \quad (1.2.6)$$

holds, then $P(s) \geqslant 0$ for all $s \in [0, T]$.

Let us now make a rough summary. Note first that in the case of $R > 0$, the optimality system (1.2.3) and the FBODE (1.2.5) are equivalent. There are, basically, three issues involved in the above: (a) Solvability of the Riccati equation; (b) Existence of optimal controls; (c) Solvability of the optimality system. From Proposition 1.2.1, we see that in general,

$$(a) \quad \Longrightarrow \quad (b) \quad \Longrightarrow \quad (c),$$

and the unique solvability of (1.2.5) for any initial pair also implies (a). Hence, roughly speaking, the above issues are equivalent.

Now, the following questions could be asked naturally:

- What if the standard condition (1.2.6) fails?
- What if there is no uniqueness of optimal controls?
- Further, what kind of results can be expected for stochastic LQ problems?

In the subsequent chapters, we shall present a systematic theory for stochastic LQ problems, including optimal control problems and two-person differential games, which answers the above questions in some satisfactory extent.

1.3   Quadratic Functionals in a Hilbert Space

Essentially, linear-quadratic optimal control problems, either deterministic or stochastic, are optimization problems for quadratic functionals in some Hilbert spaces. In this section, we will present some basic results along this line, which will play an interesting role in our later development.

Let $\mathcal{U}$ be a real Hilbert space whose inner product and norm are denoted by $\langle \cdot, \cdot \rangle$ and $\| \cdot \|$, respectively. Consider a quadratic functional $J : \mathcal{U} \to \mathbb{R}$ defined by the following:

$$J(u) = \langle Mu, u \rangle + 2\langle u, \xi \rangle,$$

where $M : \mathcal{U} \to \mathcal{U}$ is a bounded self-adjoint linear operator and $\xi \in \mathcal{U}$ is a given element.

Proposition 1.3.1 *The following hold:*

(i) *If*

$$V \equiv \inf_{u \in \mathcal{U}} J(u) > -\infty, \tag{1.3.1}$$

then $M \geqslant 0$, *which means that* $u \mapsto J(u)$ *is convex.*

(ii) *Assume* $M \geqslant 0$. *Then* J *has a minimum at* $u^* \in \mathcal{U}$ *if and only if*

$$M u^* + \xi = 0. \tag{1.3.2}$$

In this case, there exists a $v \in \mathcal{U}$ *such that*

$$u^* = -M^\dagger \xi + (I - M^\dagger M) v,$$

where $M^\dagger$ *is the pseudo-inverse of* M, *and*

$$V \equiv \inf_{u \in \mathcal{U}} J(u) = J(u^*) = -\langle M^\dagger \xi, \xi \rangle.$$

In particular, if the condition $M \geqslant 0$ *is replaced by*

$$M \geqslant \delta I \quad \text{for some } \delta > 0,$$

i.e., $u \mapsto J(u)$ *is uniformly convex, then* u^* *is uniquely given by*

$$u^* = -M^{-1} \xi.$$

Proof (i) Assume the contrary; i.e., there exists a $u \in \mathcal{U}$ such that $\langle M u, u \rangle < 0$. Then

$$J(\lambda u) = \lambda^2 \langle M u, u \rangle + 2\lambda \langle \xi, u \rangle \to -\infty, \quad \text{as } \lambda \to \infty.$$

This contradicts (1.3.1).

(ii) For any $u \in \mathcal{U}$ and $\lambda \in \mathbb{R}$, we have

$$\begin{aligned} J(u^* + \lambda u) &= \langle M(u^* + \lambda u), u^* + \lambda u \rangle + 2\langle u^* + \lambda u, \xi \rangle \\ &= J(u^*) + \lambda^2 \langle M u, u \rangle + 2\lambda \langle M u^* + \xi, u \rangle. \end{aligned}$$

So u^* is a minimum point of J if and only if

$$\lambda^2 \langle M u, u \rangle + 2\lambda \langle M u^* + \xi, u \rangle \geqslant 0, \quad \forall \lambda \in \mathbb{R}, \ \forall u \in \mathcal{U}.$$

Since $M \geqslant 0$, the above is equivalent to the following:

$$\langle M u^* + \xi, u \rangle = 0, \quad \forall u \in \mathcal{U},$$

which in turn is equivalent to (1.3.2). The rest of the proof is standard.   □

Under the necessary condition $M \geqslant 0$, we see that for any $\varepsilon > 0$,

$$J_\varepsilon(u) \triangleq J(u) + \varepsilon\|u\|^2, \quad u \in \mathcal{U}$$

is uniformly convex. Therefore, J_ε has a unique minimum point given by

$$u_\varepsilon^* = -(M + \varepsilon I)^{-1}\xi, \tag{1.3.3}$$

and

$$V_\varepsilon \equiv \inf_{u \in \mathcal{U}} J_\varepsilon(u) = -\langle (M + \varepsilon I)^{-1}\xi, \xi\rangle.$$

We have the following result.

Proposition 1.3.2 *Suppose that (1.3.1) holds. Then*

$$\lim_{\varepsilon \to 0} V_\varepsilon = V, \quad and \quad \lim_{\varepsilon \to 0} J(u_\varepsilon^*) = V. \tag{1.3.4}$$

Proof First, we note that

$$J_\varepsilon(u) = J(u) + \varepsilon\|u\|^2 \geqslant J(u) \geqslant V,$$

which leads to

$$V_\varepsilon \geqslant V. \tag{1.3.5}$$

On the other hand, for any $\delta > 0$, there exists a $u^\delta \in \mathcal{U}$ such that

$$J(u^\delta) \leqslant V + \delta.$$

Then

$$V_\varepsilon \leqslant J_\varepsilon(u^\delta) = J(u^\delta) + \varepsilon\|u^\delta\|^2 \leqslant V + \delta + \varepsilon\|u^\delta\|^2.$$

Combining this with (1.3.5), we obtain

$$V \leqslant \liminf_{\varepsilon \to 0} V_\varepsilon \leqslant \limsup_{\varepsilon \to 0} V_\varepsilon \leqslant V + \delta.$$

Since $\delta > 0$ is arbitrary, the first equality in (1.3.4) follows.

Next, since

$$V \leqslant J(u_\varepsilon^*) \leqslant J_\varepsilon(u_\varepsilon^*) = V_\varepsilon,$$

by letting $\varepsilon \to 0$, we obtain

$$V \leqslant \liminf_{\varepsilon \to 0} J(u_\varepsilon^*) \leqslant \limsup_{\varepsilon \to 0} J(u_\varepsilon^*) \leqslant \lim_{\varepsilon \to 0} V_\varepsilon = V.$$

This proves the second equality in (1.3.4). □

The second conclusion in (1.3.4) means that u_ε^* serves a minimizing family of the functional J.

Before going further, we present the following lemma.

Lemma 1.3.3 *Let* $u_k, u \in \mathcal{U}$.

(i) *If* $u_k \to u$ *weakly, then* $\|u\| \leqslant \liminf_{k\to\infty} \|u_k\|$.

(ii) $u_k \to u$ *strongly if and only if*

$$\|u_k\| \to \|u\| \quad \text{and} \quad u_k \to u \text{ weakly}.$$

Proof (i) By the Hahn-Banach theorem, we can choose a $w \in \mathcal{U}$ with $\|w\| = 1$ such that $\langle w, u \rangle = \|u\|$. Thus, using the fact that $\langle w, u_k \rangle \leqslant \|u_k\|$, we have

$$\|u\| = \langle w, u \rangle = \lim_{k\to\infty} \langle w, u_k \rangle \leqslant \liminf_{k\to\infty} \|u_k\|.$$

(ii) The necessity is obvious. Now if $\|u_k\| \to \|u\|$ and $u_k \to u$ weakly, then

$$\|u_k - u\|^2 = \|u_k\|^2 - 2\langle u, u_k \rangle + \|u\|^2 \to 0 \quad \text{as } k \to \infty.$$

This proves the sufficiency. □

Proposition 1.3.4 *Assume* $M \geqslant 0$. *Then the following are equivalent:*

(i) *the quadratic functional* J *admits a minimum point* u^*;
(ii) *the family* $\{u_\varepsilon^*\}_{\varepsilon>0}$ *is bounded;*
(iii) u_ε^* *converges strongly to a minimum point of* J *as* $\varepsilon \to 0$;
(iv) u_ε^* *converges weakly to a minimum point of* J *as* $\varepsilon \to 0$.

Proof (i) $\Rightarrow$ (ii): Let u_ε^* be defined by (1.3.3). Then,

$$V_\varepsilon = J_\varepsilon(u_\varepsilon^*) = J(u_\varepsilon^*) + \varepsilon \|u_\varepsilon^*\|^2 \geqslant V + \varepsilon \|u_\varepsilon^*\|^2,$$
$$V_\varepsilon \leqslant J_\varepsilon(u^*) = J(u^*) + \varepsilon \|u^*\|^2 = V + \varepsilon \|u^*\|^2.$$

It follows that

$$\|u_\varepsilon^*\|^2 \leqslant \frac{V_\varepsilon - V}{\varepsilon} \leqslant \|u^*\|^2, \quad \forall \varepsilon > 0. \tag{1.3.6}$$

Therefore, $\{u_\varepsilon^*\}_{\varepsilon>0}$ is bounded.

(ii) $\Rightarrow$ (iii): Because $\{u_\varepsilon^*\}_{\varepsilon>0}$ is bounded in the Hilbert space $\mathcal{U}$, there exists a sequence $\varepsilon_k > 0$ with $\lim_{k\to\infty} \varepsilon_k = 0$ such that $u_{\varepsilon_k}^*$ converges weakly to some $\bar{u} \in \mathcal{U}$. Since $u \mapsto J(u)$ is convex and continuous, it is hence sequentially weakly lower semi-continuous. Consequently, by using (1.3.4), we have

$$V \leqslant J(\bar{u}) \leqslant \liminf_{k \to \infty} J(u_{\varepsilon_k}^*) = V.$$

This shows that $\bar{u}$ is a minimum point of J. Replacing u^* by $\bar{u}$ in (1.3.6) yields

$$\|u_\varepsilon^*\|^2 \leqslant \|\bar{u}\|^2, \quad \forall \varepsilon > 0. \tag{1.3.7}$$

On the other hand, by Lemma 1.3.3(i),

$$\|\bar{u}\|^2 \leqslant \liminf_{k \to \infty} \|u_{\varepsilon_k}^*\|^2. \tag{1.3.8}$$

Combining (1.3.7) and (1.3.8), we obtain

$$\|\bar{u}\|^2 = \lim_{k \to \infty} \|u_{\varepsilon_k}^*\|^2.$$

Then we can use Lemma 1.3.3(ii) to conclude that $u_{\varepsilon_k}^* \to \bar{u}$ strongly.

To prove that $\{u_\varepsilon^*\}_{\varepsilon>0}$ itself converges strongly to $\bar{u}$ as $\varepsilon \to 0$, let us suppose that $\{u_\varepsilon^*\}_{\varepsilon>0}$ has another subsequence that converges strongly to $\hat{u}$. It suffices to show that $\hat{u} = \bar{u}$. By the previous argument, $\hat{u}$ is another minimum point of J. Thus, making use of the convexity of J, we see that $\frac{1}{2}(\bar{u} + \hat{u})$ is also a minimum point of J. Now replacing u^* in (1.3.6) by $\frac{1}{2}(\bar{u} + \hat{u})$, we obtain

$$\|u_\varepsilon^*\|^2 \leqslant \left\|\frac{\bar{u} + \hat{u}}{2}\right\|^2,$$

which leads to

$$\|\bar{u}\|^2 \leqslant \left\|\frac{\bar{u} + \hat{u}}{2}\right\|^2, \quad \|\hat{u}\|^2 \leqslant \left\|\frac{\bar{u} + \hat{u}}{2}\right\|^2.$$

Adding the above two inequalities yields

$$2\|\bar{u}\|^2 + 2\|\hat{u}\|^2 \leqslant \|\bar{u} + \hat{u}\|^2 = \|\bar{u}\|^2 + \|\hat{u}\|^2 + 2\langle \bar{u}, \hat{u} \rangle,$$

from which it follows that $\|\bar{u} - \hat{u}\|^2 \leqslant 0$. Therefore, $\bar{u} = \hat{u}$.

The implications (iii) $\Rightarrow$ (iv) and (iv) $\Rightarrow$ (i) are trivial. $\square$

Chapter 2
Linear-Quadratic Optimal Controls in Finite Horizons

Abstract This chapter is devoted to a study of stochastic linear-quadratic optimal control problems in a finite horizon from two points of view: open-loop and closed-loop solvabilities. A simple example shows that these two solvabilities are essentially different. Open-loop solvability is established by studying the solvability of a constrained linear forward-backward stochastic differential equation. Closed-loop solvability is reduced to the existence of a regular solution to the associated differential Riccati equation, which is implied by the uniform convexity of the quadratic cost functional. The relation between open-loop and closed-loop solvabilities, as well as some other aspects, such as conditions ensuring the convexity of the cost functional, finiteness of the problem and construction of minimizing sequences, are also discussed.

Keywords Linear-quadratic optimal control · Finite horizon · Open-loop solvability · Closed-loop solvability · Differential Riccati equation · Uniform convexity · Finiteness

Throughout this book, we let $(\Omega, \mathcal{F}, \mathbb{F}, \mathbb{P})$ be a complete filtered probability space on which a standard one-dimensional Brownian motion $W = \{W(t); 0 \leqslant t < \infty\}$ is defined.[1] We assume that $\mathbb{F} = \{\mathcal{F}_t\}_{t \geqslant 0}$ is the natural filtration of W augmented by all the $\mathbb{P}$-null sets in $\mathcal{F}$. Hence, $\mathbb{F}$ automatically satisfies the *usual conditions*.

Let $T > 0$ be a fixed time horizon. For any $t \in [0, T)$ and Euclidean space $\mathbb{H}$, we denote by $L^p(t, T; \mathbb{H})$ $(1 \leqslant p < \infty)$ the space of all $\mathbb{H}$-valued functions that are pth power Lebesgue integrable on $[t, T]$, by $L^\infty(t, T; \mathbb{H})$ the space of all Lebesgue measurable, essentially bounded $\mathbb{H}$-valued functions on $[t, T]$, and by $C([t, T]; \mathbb{H})$ the space of all $\mathbb{H}$-valued continuous functions on $[t, T]$. For spaces of random variables and stochastic processes, we employ the following notation:

---

[1]Multi-dimensional Brownian motion case can be treated similarly.

$$L^2_{\mathcal{F}_t}(\Omega; \mathbb{H}) = \left\{ \xi : \Omega \to \mathbb{H} \mid \xi \text{ is } \mathcal{F}_t\text{-measurable, and} \mathbb{E}|\xi|^2 < \infty \right\},$$

$$L^2_{\mathbb{F}}(t, T; \mathbb{H}) = \left\{ \varphi : [t, T] \times \Omega \to \mathbb{H} \mid \varphi \text{ is } \mathbb{F}\text{-progressively} \right.$$

$$\left. \text{measurable, and } \mathbb{E} \int_t^T |\varphi(s)|^2 ds < \infty \right\},$$

$$L^2_{\mathbb{F}}(\Omega; C([t, T]; \mathbb{H})) = \left\{ \varphi : [t, T] \times \Omega \to \mathbb{H} \mid \varphi \text{ is } \mathbb{F}\text{-adapted, continuous,} \right.$$

$$\left. \text{and } \mathbb{E}\left[\sup_{t \leqslant s \leqslant T} |\varphi(s)|^2 \right] < \infty \right\},$$

$$L^2_{\mathbb{F}}(\Omega; L^1(t, T; \mathbb{H})) = \left\{ \varphi : [t, T] \times \Omega \to \mathbb{H} \mid \varphi \text{ is } \mathbb{F}\text{-progressively} \right.$$

$$\left. \text{measurable, and } \mathbb{E}\left(\int_t^T |\varphi(s)| ds \right)^2 < \infty \right\}.$$

When $\mathbb{H} = \mathbb{R}^n$, we simply write

$$\mathcal{X}[t, T] = L^2_{\mathbb{F}}(\Omega; C([t, T]; \mathbb{R}^n)), \quad \mathcal{X}_t = L^2_{\mathcal{F}_t}(\Omega; \mathbb{R}^n).$$

We shall impose the L^2-inner products on $L^2_{\mathcal{F}_t}(\Omega; \mathbb{H})$ and $L^2_{\mathbb{F}}(t, T; \mathbb{H})$ to make them into Hilbert spaces. For notational simplicity, we will use $\langle \cdot, \cdot \rangle$ to denote all inner products in different Hilbert spaces which can be identified from the context. With the norm

$$\|\varphi\| = \left[\mathbb{E}\left(\sup_{t \leqslant s \leqslant T} |\varphi(s)|^2 \right) \right]^{\frac{1}{2}},$$

$L^2_{\mathbb{F}}(\Omega; C([t, T]; \mathbb{H}))$ is a Banach space, and its dual space is $L^2_{\mathbb{F}}(\Omega; \mathcal{M}(t, T; \mathbb{H}))$, where $\mathcal{M}(t, T; \mathbb{H})$ denotes the space of $\mathbb{H}$-valued (Radon) measures on $[t, T]$. For each $f \in L^1(t, T; \mathbb{H})$, we can define a measure μ on $[t, T]$ by

$$\mu(E) = \int_E f(s) ds; \quad E \subseteq \mathcal{B}[t, T],$$

where $\mathcal{B}[t, T]$ is the Borel σ-field of $[t, T]$. In this sense, $L^2_{\mathbb{F}}(\Omega; L^1(t, T; \mathbb{H}))$ is a subspace of $L^2_{\mathbb{F}}(\Omega; \mathcal{M}(t, T; \mathbb{H}))$.

Let $\mathbb{S}^n$ be the space of all symmetric $n \times n$ real matrices. For an $\mathbb{S}^n$-valued function F on $[t, T]$, by $F \geqslant 0$, we mean that $F(s)$ is positive semi-definite (or non-negative) for almost all $s \in [t, T]$, and by $F \gg 0$, we mean that F is uniformly positive definite, i.e., there exists a constant $\delta > 0$ such that

$$F(s) \geqslant \delta I_n, \quad \text{a.e. } s \in [t, T],$$

where I_n is the identity matrix of size n.

2.1   Formulation of the Problem

Consider the controlled linear SDE on a finite time horizon $[t, T]$:

$$\begin{cases} dX(s) = [A(s)X(s) + B(s)u(s) + b(s)]ds \\ \qquad\qquad + [C(s)X(s) + D(s)u(s) + \sigma(s)]dW(s), \\ X(t) = x, \end{cases} \qquad (2.1.1)$$

where A, B, C, and D are given deterministic matrix-valued functions of proper dimensions, called the *coefficients* of the *state equation* (2.1.1), and b, σ are given vector-valued $\mathbb{F}$-progressively measurable processes, called the *nonhomogeneous terms*. In the above, the solution X of (2.1.1) is called the *state process*, and u is called the *control process*. In this chapter, we assume the following standard assumption.

(H1) The coefficients and the nonhomogeneous terms of (2.1.1) satisfy

$$\begin{cases} A \in L^1(0, T; \mathbb{R}^{n \times n}), & B \in L^2(0, T; \mathbb{R}^{n \times m}), & b \in L^2_{\mathbb{F}}(\Omega; L^1(0, T; \mathbb{R}^n)), \\ C \in L^2(0, T; \mathbb{R}^{n \times n}), & D \in L^{\infty}(0, T; \mathbb{R}^{n \times m}), & \sigma \in L^2_{\mathbb{F}}(0, T; \mathbb{R}^n). \end{cases}$$

Let $\mathcal{U}[t, T] = L^2_{\mathbb{F}}(t, T; \mathbb{R}^m)$. We call any element $u \in \mathcal{U}[t, T]$ an *admissible control* (on $[t, T]$). Now we may state the basic existence and uniqueness result for our state equation (2.1.1).

Proposition 2.1.1 *Let (H1) hold. Then for any initial pair* $(t, x) \in [0, T) \times \mathbb{R}^n$ *and admissible control* $u \in \mathcal{U}[t, T]$, *the state equation (2.1.1) admits a unique solution* $X(\cdot) \equiv X(\cdot\,; t, x, u) \in \mathcal{X}[t, T]$. *Moreover, there exists a constant* $K > 0$, *independent of* (t, x, u), *such that*

$$\mathbb{E}\left[\sup_{t \leqslant s \leqslant T} |X(s)|^2\right] \leqslant K\mathbb{E}\left[|x|^2 + \left(\int_t^T |b(s)|ds\right)^2 + \int_t^T |\sigma(s)|^2 ds \right.$$
$$\left. + \int_t^T |u(s)|^2 ds\right].$$

To measure the performance of the control process u, we introduce the following *cost functional*:

$$J(t, x; u) = \mathbb{E}\Bigg\{\langle GX(T), X(T)\rangle + 2\langle g, X(T)\rangle$$
$$+ \int_t^T \left[\left\langle \begin{pmatrix} Q(s) & S(s)^{\top} \\ S(s) & R(s) \end{pmatrix} \begin{pmatrix} X(s) \\ u(s) \end{pmatrix}, \begin{pmatrix} X(s) \\ u(s) \end{pmatrix}\right\rangle\right.$$
$$\left.+ 2\left\langle \begin{pmatrix} q(s) \\ \rho(s) \end{pmatrix}, \begin{pmatrix} X(s) \\ u(s) \end{pmatrix}\right\rangle\right]ds\Bigg\}. \qquad (2.1.2)$$

We introduce the following assumption for the weighting coefficients.

(H2) The weighting coefficients in the cost functional satisfy

$$\begin{cases} G \in \mathbb{S}^n, \ g \in L^2_{\mathcal{F}_T}(\Omega; \mathbb{R}^n), \quad Q \in L^1(0, T; \mathbb{S}^n), \quad S \in L^2(0, T; \mathbb{R}^{m \times n}), \\ q \in L^2_{\mathbb{F}}(\Omega; L^1(0, T; \mathbb{R}^n)), \quad \rho \in L^2_{\mathbb{F}}(0, T; \mathbb{R}^m), \quad R \in L^\infty(0, T; \mathbb{S}^m). \end{cases}$$

Clearly, under assumptions (H1)–(H2), for any initial pair $(t, x) \in [0, T) \times \mathbb{R}^n$ and admissible control $u \in \mathcal{U}[t, T]$, the cost functional (2.1.2) is well-defined. We point out that (H2) does not impose any positive-definiteness/non-negativeness conditions on Q, R, or G. Now, we pose the following problem.

Problem (SLQ). For any given initial pair $(t, x) \in [0, T) \times \mathbb{R}^n$, find an admissible control $\bar{u} \in \mathcal{U}[t, T]$ such that

$$J(t, x; \bar{u}) = \inf_{u \in \mathcal{U}[t,T]} J(t, x; u) \triangleq V(t, x).$$

The above is called a *stochastic linear-quadratic optimal control problem* (SLQ problem, for short), and $V(t, x)$ is called the *value function* of Problem (SLQ). When $b, \sigma, g, q, \rho = 0$, we denote the corresponding Problem (SLQ) by Problem (SLQ)^0. The corresponding cost functional and value function are denoted by $J^0(t, x; u)$ and $V^0(t, x)$, respectively.

Definition 2.1.2 Problem (SLQ) is said to be

   (i) *finite at initial pair* $(t, x) \in [0, T] \times \mathbb{R}^n$ if $V(t, x) > -\infty$;
  (ii) *finite at* $t \in [0, T]$ if $V(t, x) > -\infty$ for all $x \in \mathbb{R}^n$;
 (iii) *finite* if $V(t, x) > -\infty$ for all $x \in \mathbb{R}^n$ and all $t \in [0, T]$.

Definition 2.1.3 Problem (SLQ) is said to be

   (i) *(uniquely) open-loop solvable at* $(t, x) \in [0, T) \times \mathbb{R}^n$ if there exists a (unique) $\bar{u}(\cdot) \equiv \bar{u}(\cdot\,; t, x) \in \mathcal{U}[t, T]$ (depending on (t, x)) such that

$$J(t, x; \bar{u}) \leqslant J(t, x; u), \quad \forall u \in \mathcal{U}[t, T];$$

Such a $\bar{u}$ is called an *open-loop optimal control for* (t, x).
  (ii) *(uniquely) open-loop solvable at* t if it is (uniquely) open-loop solvable at (t, x) for all $x \in \mathbb{R}^n$;
 (iii) *(uniquely) open-loop solvable* if it is (uniquely) open-loop solvable at any initial pair $(t, x) \in [0, T) \times \mathbb{R}^n$.

Next let $\boldsymbol{\Theta}[t, T] = L^2(t, T; \mathbb{R}^{m \times n})$. For any pair $(\Theta, v) \in \boldsymbol{\Theta}[t, T] \times \mathcal{U}[t, T]$, which is called a *closed-loop strategy*, we consider the SDE

$$\begin{cases} dX(s) = [(A + B\Theta)X + Bv + b]ds \\ \qquad\qquad + [(C + D\Theta)X + Dv + \sigma]dW(s), \quad s \in [t, T], \qquad (2.1.3) \\ X(t) = x, \end{cases}$$

where, for convenience, we have suppressed the variable s in the drift and diffusion terms. Equation (2.1.3) is called a *closed-loop system* under the closed-loop strategy (Θ, v) (with the initial pair (t, x)). Clearly, under (H1), it admits a unique solution, which is denoted by $X(\cdot) \equiv X(\cdot; t, x, \Theta, v)$. Notice that $u \triangleq \Theta X + v$ is an admissible control and that

$$
J(t, x; \Theta X + v) = \mathbb{E}\bigg\{ \langle G X(T), X(T) \rangle + 2 \langle g, X(T) \rangle
$$

$$
+ \int_t^T \bigg[\bigg\langle \begin{pmatrix} Q & S^\top \\ S & R \end{pmatrix} \begin{pmatrix} X \\ \Theta X + v \end{pmatrix}, \begin{pmatrix} X \\ \Theta X + v \end{pmatrix} \bigg\rangle
$$

$$
+ 2 \bigg\langle \begin{pmatrix} q \\ \rho \end{pmatrix}, \begin{pmatrix} X \\ \Theta X + v \end{pmatrix} \bigg\rangle \bigg] ds \bigg\}.
$$

Definition 2.1.4 A pair $(\bar{\Theta}, \bar{v}) \in \Theta[t, T] \times \mathcal{U}[t, T]$ is called a *closed-loop optimal strategy* of Problem (SLQ) on $[t, T]$ if

$$
\begin{aligned}
J(t, x; \bar{\Theta} \bar{X} + \bar{v}) &\leqslant J(t, x; \Theta X + v), \\
\forall x \in \mathbb{R}^n, \ \forall (\Theta, v) &\in \Theta[t, T] \times \mathcal{U}[t, T],
\end{aligned}
\tag{2.1.4}
$$

where $\bar{X}$ is the strong solution to the closed-loop system (2.1.3) with $(\Theta, v) = (\bar{\Theta}, \bar{v})$. If a closed-loop optimal strategy (uniquely) exists on $[t, T]$, we say Problem (SLQ) is *(uniquely) closed-loop solvable on $[t, T]$*. We simply say Problem (SLQ) is *(uniquely) closed-loop solvable* if it is (uniquely) closed-loop solvable on any $[t, T]$.

We point out that in the definition of closed-loop optimal strategy, (2.1.4) has to be true for all $x \in \mathbb{R}^n$. The following result provides some equivalent definitions of closed-loop optimal strategy.

Proposition 2.1.5 *Let (H1) hold, and let $(\bar{\Theta}, \bar{v}) \in \Theta[t, T] \times \mathcal{U}[t, T]$. The following are equivalent:*

(i) $(\bar{\Theta}, \bar{v})$ *is a closed-loop optimal strategy of Problem (SLQ) on $[t, T]$;*
(ii) for any $x \in \mathbb{R}^n$ and $v \in \mathcal{U}[t, T]$,

$$
J(t, x; \bar{\Theta} \bar{X} + \bar{v}) \leqslant J(t, x; \bar{\Theta} X + v); \tag{2.1.5}
$$

(iii) for any $x \in \mathbb{R}^n$ and $u \in \mathcal{U}[t, T]$,

$$
J(t, x; \bar{\Theta} \bar{X} + \bar{v}) \leqslant J(t, x; u). \tag{2.1.6}
$$

Proof By taking $(\Theta, v) = (\bar{\Theta}, v)$ in (2.1.4) we obtain (i) $\Rightarrow$ (ii).

To prove the implication (ii) $\Rightarrow$ (iii), fix an arbitrary $x \in \mathbb{R}^n$ and a $u \in \mathcal{U}[t, T]$, and let X be the corresponding state process. We may rewrite the first equation in (2.1.1) as

$$dX(s) = [(A + B\bar{\Theta})X + B(u - \bar{\Theta}X) + b]ds$$
$$+ [(C + D\bar{\Theta})X + D(u - \bar{\Theta}X) + \sigma]dW(s).$$

Then, with $v = u - \bar{\Theta}X$, we have

$$J(t, x; u) = J(t, x; \bar{\Theta}X + v),$$

and (2.1.6) follows from (2.1.5).

Finally, we establish (iii) $\Rightarrow$ (i). For any $(\Theta, v) \in \mathbf{\Theta}[t, T] \times \mathcal{U}[t, T]$, let X be the state process under this closed-loop strategy (with the initial pair (t, x)). Then by setting $u = \Theta X + v$, we obtain (2.1.4) from (2.1.6).   □

From the above result, one sees that if $(\bar{\Theta}, \bar{v})$ is a closed-loop optimal strategy of Problem (SLQ) on $[t, T]$, then the *outcome* $\bar{u} \equiv \bar{\Theta}\bar{X} + \bar{v}$ is an open-loop optimal control of Problem (SLQ) for the initial pair $(t, \bar{X}(t))$. Hence, the existence of closed-loop optimal strategies implies the existence of open-loop optimal controls. But the reverse implication is not necessarily true. Here is such an example.

Example 2.1.6 Consider the following Problem (SLQ)^0 with one-dimensional state equation

$$\begin{cases} dX(s) = [u_1(s) + u_2(s)]ds + [u_1(s) - u_2(s)]dW(s), & s \in [t, 1], \\ X(t) = x, \end{cases}$$

and cost functional

$$J^0(t, x; u) = \mathbb{E}[X(1)^2].$$

In this example, $u = (u_1, u_2)^\top$ is the control process. Obviously,

$$V^0(t, x) = \inf_{u \in \mathcal{U}[t,T]} J^0(t, x; u) \geqslant 0, \quad \forall (t, x) \in [0, 1] \times \mathbb{R}.$$

On the other hand, for any $(t, x) \in [0, 1) \times \mathbb{R}$, the state process corresponding to the control

$$u^{\beta,x}(s) = -\frac{\beta x}{2} \mathbf{1}_{[t,t+\frac{1}{\beta}]}(s) \begin{pmatrix} 1 \\ 1 \end{pmatrix}, \quad s \in [t, 1],$$

where $\beta \geqslant \frac{1}{1-t}$, satisfies

$$X(s) = 0, \quad s \in [t + 1/\beta, 1].$$

Hence, $J^0(t, x; u^{\beta,x}) = 0$. This shows that $\{u^{\beta,x}(\cdot) : \beta \geqslant \frac{1}{1-t}\}$ is a family of open-loop optimal controls for the initial pair (t, x), and therefore,

$$V^0(t, x) = \begin{cases} 0, & t \in [0, 1), \\ x^2, & t = 1, \end{cases}$$

which is discontinuous at $t = 1$, $x \neq 0$. Note also that if we take $\beta = \frac{1}{1-t}$, then the corresponding open-loop optimal control, denoted by $\bar{u}$, is given by

$$\bar{u}(s) = -\frac{x}{2(1-t)} \begin{pmatrix} 1 \\ 1 \end{pmatrix}, \quad s \in [t, 1],$$

which is a constant vector (depending only on the initial pair (t, x)) and hence continuous (in $s \in [t, 1]$). Now, we claim that this problem is not closed-loop solvable on any $[t, 1]$, with $t \in [0, 1)$. In fact, if for some $t \in [0, 1)$, there exists a closed-loop optimal strategy

$$\bar{\Theta} = \begin{pmatrix} \bar{\Theta}_1 \\ \bar{\Theta}_2 \end{pmatrix}, \quad \bar{v} = \begin{pmatrix} \bar{v}_1 \\ \bar{v}_2 \end{pmatrix},$$

then, by Proposition 2.1.5, we have

$$0 \leqslant J^0(t, x; \bar{\Theta}\bar{X} + \bar{v}) \leqslant J^0(t, x; u^{\beta, x}) = 0, \quad \forall x \in \mathbb{R}.$$

This implies that for any $x \in \mathbb{R}$, the solution $\bar{X}$ to the closed-loop system

$$\begin{cases} d\bar{X}(s) = \{[\bar{\Theta}_1(s) + \bar{\Theta}_2(s)]\bar{X}(s) + [\bar{v}_1(s) + \bar{v}_2(s)]\}ds \\ \qquad\qquad + \{[\bar{\Theta}_1(s) - \bar{\Theta}_2(s)]\bar{X}(s) + [\bar{v}_1(s) - \bar{v}_2(s)]\}dW(s), \\ \bar{X}(t) = x, \end{cases}$$

satisfies $\bar{X}(1) = 0$. Taking expectation in the above leads to

$$\begin{cases} d[\mathbb{E}\bar{X}(s)] = \{[\bar{\Theta}_1(s) + \bar{\Theta}_2(s)]\mathbb{E}\bar{X}(s) + \mathbb{E}[\bar{v}_1(s) + \bar{v}_2(s)]\}ds, \\ \mathbb{E}\bar{X}(t) = x. \end{cases}$$

Consequently,

$$0 = \mathbb{E}\bar{X}(1) = e^{\int_t^1 [\bar{\Theta}_1(s) + \bar{\Theta}_2(s)]ds} x + \int_t^1 e^{\int_r^1 [\bar{\Theta}_1(s) + \bar{\Theta}_2(s)]ds} \mathbb{E}[\bar{v}_1(r) + \bar{v}_2(r)]dr.$$

Since $(\bar{\Theta}, \bar{v})$ is required to be independent of $x \in \mathbb{R}$, the above cannot be true for all x. This contradiction shows that the problem is not closed-loop solvable on any $[t, 1]$.

Due to the above situation, we distinguish the notions of open-loop and closed-loop solvability for Problem (SLQ). Here we emphasize again that for given initial time $t \in [0, T)$, an open-loop optimal control is allowed to depend on the initial state x, whereas a closed-loop optimal strategy is required to be independent of x.

2.2   Representation of the Cost Functional

In this section, we will give a representation of our quadratic cost functional $J(t, x; u)$ in the Hilbert space $\mathcal{U}[t, T]$, and then apply the results from Chap. 1, Sect. 1.3, to obtain basic properties of $J(t, x; u)$.

Let $\Phi = \{\Phi(s); 0 \leqslant s \leqslant T\}$ be the solution to the linear matrix SDE

$$\begin{cases} d\Phi(s) = A(s)\Phi(s)ds + C(s)\Phi(s)dW(s), & s \in [0, T], \\ \Phi(0) = I_n. \end{cases} \tag{2.2.1}$$

Then the process Φ is invertible, and its inverse $\Phi(s)^{-1}$ satisfies the following SDE:

$$\begin{cases} d[\Phi(s)^{-1}] = -[\Phi(s)^{-1}][A(s) - C(s)^2]ds - [\Phi(s)^{-1}]C(s)dW(s), \\ \Phi(0)^{-1} = I_n. \end{cases}$$

By the variation of constants formula, the solution $X_x^u(\cdot) \equiv X(\cdot\,; t, x, u)$ to the state equation (2.1.1) can be written as

$$X_x^u(s) = (\Gamma_t x)(s) + (L_t u)(s) + h_t(s), \quad s \in [t, T],$$

where, for $s \in [t, T]$,

$$(\Gamma_t x)(s) = \Phi(s)\Phi(t)^{-1}x,$$

$$(L_t u)(s) = \Phi(s)\int_t^s \Phi(r)^{-1}[B(r) - C(r)D(r)]u(r)dr$$

$$\qquad\qquad + \Phi(s)\int_t^s \Phi(r)^{-1}D(r)u(r)dW(r),$$

$$h_t(s) = \Phi(s)\int_t^s \Phi(r)^{-1}[b(r) - C(r)\sigma(r)]dr + \Phi(s)\int_t^s \Phi(r)^{-1}\sigma(r)dW(r).$$

So with

$$\widehat{\Gamma}_t x \triangleq (\Gamma_t x)(T), \quad \widehat{L}_t u \triangleq (L_t u)(T),$$

$X_x^u(T)$ can be rewritten as

$$X_x^u(T) = \widehat{\Gamma}_t x + \widehat{L}_t u + h_t(T).$$

We observe that the process

$$\overset{\circ}{X}{}_x^u(s) \triangleq (\Gamma_t x)(s) + (L_t u)(s), \quad s \in [t, T]$$

is the solution to the state equation (2.1.1) with $b = \sigma = 0$. It is easily seen that the operators

$$L_t : \mathcal{U}[t, T] \to \mathcal{X}[t, T], \qquad \Gamma_t : \mathbb{R}^n \to \mathcal{X}[t, T],$$
$$\widehat{L}_t : \mathcal{U}[t, T] \to \mathcal{X}_T, \qquad \widehat{\Gamma}_t : \mathbb{R}^n \to \mathcal{X}_T,$$

are all linear and bounded. Now let L_t^*, Γ_t^*, $\widehat{L}_t^*$, and $\widehat{\Gamma}_t^*$ denote the adjoint operators of L_t, Γ_t, $\widehat{L}_t$, and $\widehat{\Gamma}_t$, respectively. If we regard the weighting matrices in the cost functional as appropriate operators, then the cost functional can be represented as

$$
\begin{aligned}
J(t, x; u) = \ &\langle M_2(t)u, u \rangle + 2\langle M_1(t)x, u \rangle + \langle M_0(t)x, x \rangle \\
&+ 2\langle x, y_t \rangle + 2\langle u, \nu_t \rangle + c_t,
\end{aligned}
\tag{2.2.2}
$$

where

$$
\begin{cases}
M_2(t) = \widehat{L}_t^* G \widehat{L}_t + L_t^* Q L_t + S L_t + L_t^* S^\top + R, \\
M_1(t) = \widehat{L}_t^* G \widehat{\Gamma}_t + L_t^* Q \Gamma_t + S \Gamma_t, \\
M_0(t) = \widehat{\Gamma}_t^* G \widehat{\Gamma}_t + \Gamma_t^* Q \Gamma_t, \\
y_t = \widehat{\Gamma}_t^* [G h_t(T) + g] + \Gamma_t^*(Q h_t + q), \\
\nu_t = \widehat{L}_t^* [G h_t(T) + g] + L_t^*(Q h_t + q) + S h_t + \rho, \\
c_t = \langle G h_t(T), h_t(T) \rangle + 2\langle g, h_t(T) \rangle + \langle Q h_t, h_t \rangle + 2\langle q, h_t \rangle.
\end{cases}
\tag{2.2.3}
$$

Note that for given $t \in [0, T)$,

- $M_2(t)$ is a bounded self-adjoint operator from $\mathcal{U}[t, T]$ into itself;
- $M_1(t)$ is bounded operator from $\mathbb{R}^n$ into $\mathcal{U}[t, T]$;
- $M_0(t)$ is an $n \times n$ symmetric matrix, i.e, $M_0(t) \in \mathbb{S}^n$;
- $y_t \in \mathbb{R}^n$, $\nu_t \in \mathcal{U}[t, T]$, and $c_t \in \mathbb{R}$;
- $\nu_t, y_t, c_t = 0$ when $b, \sigma, g, q, \rho = 0$ (noting that $b, \sigma = 0 \Rightarrow h_t = 0$).

Clearly, (2.2.2) can also be written as

$$J(t, x; u) = \langle M_2(t)u, u \rangle + \langle \mathcal{D}_u J(t, x; 0), u \rangle + J(t, x; 0),$$

where $\mathcal{D}_u J(t, x; u)$ is the Fréchet derivative of the mapping $u \mapsto J(t, x; u)$. This is actually the Taylor expansion for the quadratic functional $u \mapsto J(t, x; u)$ about $u = 0$. Further, it follows from the (2.2.2) that

$$J^0(t, x; u) = \langle M_2(t)u, u \rangle + 2\langle M_1(t)x, u \rangle + \langle M_0(t)x, x \rangle. \tag{2.2.4}$$

We see that the operators $M_2(t)$, $M_1(t)$, and $M_0(t)$ defined in (2.2.3) seem to be complicated, due to the appearance of L_t^*, $\widehat{L}_t^*$, Γ_t^*, and $\widehat{\Gamma}_t^*$. The following result gives representations of these operators in terms of solutions to some SDEs. Notice that we will frequently suppress the variable s for notational simplicity.

Proposition 2.2.1 *Let (H1)–(H2) hold. For any $x \in \mathbb{R}^n$ and $u \in \mathcal{U}[t, T]$, let $(\mathring{X}^u_x, \mathring{Y}^u_x, \mathring{Z}^u_x)$ be the adapted solution to the following (decoupled) linear forward-backward stochastic differential equation (FBSDE, for short) on $[t, T]$:*

$$\begin{cases} d\mathring{X}^u_x(s) = (A\mathring{X}^u_x + Bu)ds + (C\mathring{X}^u_x + Du)dW(s), \\ d\mathring{Y}^u_x(s) = -(A^\top \mathring{Y}^u_x + C^\top \mathring{Z}^u_x + Q\mathring{X}^u_x + S^\top u)ds + \mathring{Z}^u_x dW(s), \\ \mathring{X}^u_x(t) = x, \quad \mathring{Y}^u_x(T) = G\mathring{X}^u_x(T). \end{cases} \quad (2.2.5)$$

Then

$$[M_2(t)u](s) = B(s)^\top \mathring{Y}^u_0(s) + D(s)^\top \mathring{Z}^u_0(s) + S(s)\mathring{X}^u_0(s) + R(s)u(s),$$
$$[M_1(t)x](s) = B(s)^\top \mathring{Y}^0_x(s) + D(s)^\top \mathring{Z}^0_x(s) + S(s)\mathring{X}^0_x(s),$$
$$M_0(t)x = \mathbb{E}[\mathring{Y}^0_x(t)].$$

Moreover, M_0 solves the Lyapunov equation

$$\begin{cases} \dot{M}_0(t) + M_0(t)A(t) + A(t)^\top M_0(t) \\ \qquad + C(t)^\top M_0(t)C(t) + Q(t) = 0, \quad t \in [0, T], \\ M_0(T) = G, \end{cases} \quad (2.2.6)$$

and admits the following representation:

$$M_0(t) = \mathbb{E}\Big\{ \big[\Phi(T)\Phi(t)^{-1}\big]^\top G\big[\Phi(T)\Phi(t)^{-1}\big]$$
$$+ \int_t^T \big[\Phi(s)\Phi(t)^{-1}\big]^\top Q(s)\big[\Phi(s)\Phi(t)^{-1}\big]ds \Big\}, \quad (2.2.7)$$

where Φ is the solution of (2.2.1).

Proof Let us first identify the adjoint operators L_t^*, Γ_t^*, $\widehat{L}_t^*$, and $\widehat{\Gamma}_t^*$. To this end, we let $(\mathcal{Y}^\xi_\eta, \mathcal{Z}^\xi_\eta)$ be the adapted solution to the following backward stochastic differential equation (BSDE, for short):

$$\begin{cases} d\mathcal{Y}^\xi_\eta(s) = -(A^\top \mathcal{Y}^\xi_\eta + C^\top \mathcal{Z}^\xi_\eta + \xi)ds + \mathcal{Z}^\xi_\eta dW(s), \quad s \in [t, T], \\ \mathcal{Y}^\xi_\eta(T) = \eta, \end{cases}$$

where $\xi \in L^2_{\mathbb{F}}(\Omega; L^1(t, T; \mathbb{R}^n))$ and $\eta \in \mathcal{X}_T$. Applying Itô's formula to $s \mapsto \langle \mathring{X}^u_x(s), Y^\xi_\eta(s)\rangle$, we obtain

$$\mathbb{E}\big[\langle \mathring{X}^u_x(T), \eta\rangle - \langle x, \mathcal{Y}^\xi_\eta(t)\rangle\big]$$

$$= \mathbb{E}\int_t^T \Big[\langle A\mathring{X}^u_x + Bu, \mathcal{Y}^\xi_\eta\rangle - \langle \mathring{X}^u_x, A^\top\mathcal{Y}^\xi_\eta + C^\top\mathcal{Z}^\xi_\eta + \xi\rangle$$

$$+ \langle C\mathring{X}^u_x + Du, \mathcal{Z}^\xi_\eta\rangle\Big]ds$$

$$= \mathbb{E}\int_t^T \Big[\langle u, B^\top\mathcal{Y}^\xi_\eta + D^\top\mathcal{Z}^\xi_\eta\rangle - \langle X^u_x, \xi\rangle\Big]ds,$$

which is equivalent to

$$\langle \widehat{L}_t u + \widehat{\Gamma}_t x, \eta\rangle - \langle x, \mathbb{E}\mathcal{Y}^\xi_0(t) + \mathbb{E}\mathcal{Y}^0_\eta(t)\rangle$$

$$= \langle u, B^\top\mathcal{Y}^\xi_0 + D^\top\mathcal{Z}^\xi_0\rangle + \langle u, B^\top\mathcal{Y}^0_\eta + D^\top\mathcal{Z}^0_\eta\rangle - \langle L_t u + \Gamma_t x, \xi\rangle.$$

From the above one can deduce

$$(L_t^*\xi)(s) = B(s)^\top\mathcal{Y}^\xi_0(s) + D(s)^\top\mathcal{Z}^\xi_0(s), \quad \Gamma_t^*\xi = \mathbb{E}\mathcal{Y}^\xi_0(t), \qquad (2.2.8)$$

$$(\widehat{L}_t^*\eta)(s) = B(s)^\top\mathcal{Y}^0_\eta(s) + D(s)^\top\mathcal{Z}^0_\eta(s), \quad \widehat{\Gamma}_t^*\eta = \mathbb{E}\mathcal{Y}^0_\eta(t), \qquad (2.2.9)$$

by considering the following four cases:

$$\begin{array}{llll} x = 0, & \eta = 0, & (\text{get } L_t^*); & x = 0, \quad \xi = 0, \quad (\text{get } \widehat{L}_t^*); \\ u = 0, & \eta = 0, & (\text{get } \Gamma_t^*); & u = 0, \quad \xi = 0, \quad (\text{get } \widehat{\Gamma}_t^*). \end{array}$$

From (2.2.8) and (2.2.9), it follows by linearity that

$$\widehat{L}_t^*\eta + L_t^*\xi = B^\top\mathcal{Y}^\xi_\eta + D^\top\mathcal{Z}^\xi_\eta, \quad \widehat{\Gamma}_t^*\eta + \Gamma_t^*\xi = \mathbb{E}\mathcal{Y}^\xi_\eta(t).$$

Now we take

$$\xi = Q\mathring{X}^u_0 + S^\top u, \quad \eta = G\mathring{X}^u_0(T),$$

then $(\mathcal{Y}^\xi_\eta, \mathcal{Z}^\xi_\eta)$ coincides with the adapted solution $(\mathring{Y}^u_0, \mathring{Z}^u_0)$ to the BSDE in (2.2.5) with $x = 0$. Thus,

$$M_2(t)u = \widehat{L}_t^* G\widehat{L}_t u + L_t^* QL_t u + SL_t u + L_t^* S^\top u + Ru$$

$$= \widehat{L}_t^* G\mathring{X}^u_0(T) + L_t^*(Q\mathring{X}^u_0 + S^\top u) + S\mathring{X}^u_0 + Ru$$

$$= B^\top\mathring{Y}^u_0 + D^\top\mathring{Z}^u_0 + S\mathring{X}^u_0 + Ru,$$

which proves (i). If we take

$$\xi = Q\mathring{X}^0_x, \quad \eta = G\mathring{X}^0_x(T),$$

then $(\mathcal{Y}_\eta^\xi, Z_\eta^\xi)$ coincides with the adapted solution $(\mathring{Y}_x^0, \mathring{Z}_x^0)$ to the BSDE in (2.2.5) with $u = 0$. Thus,

$$
\begin{aligned}
M_1(t)x &= \widehat{L}_t^* G \widehat{\Gamma}_t x + L_t^* Q \Gamma_t x + S \Gamma_t x = \widehat{L}_t^* G \mathring{X}_x^0(T) + L_t^* Q \mathring{X}_x^0 + S \mathring{X}_x^0 \\
&= B^\top \mathring{Y}_x^0 + D^\top \mathring{Z}_x^0 + S \mathring{X}_x^0, \\
M_0(t)x &= \widehat{\Gamma}_t^* G \widehat{\Gamma}_t x + \Gamma_t^* Q \Gamma_t x = \widehat{\Gamma}_t^* G \mathring{X}_x^0(T) + \Gamma_t^* Q \mathring{X}_x^0 = \mathbb{E} \mathring{Y}_x^0(t),
\end{aligned}
$$

which proves (ii).

Finally, we have from (2.2.2) that for any $x \in \mathbb{R}^n$,

$$
\begin{aligned}
\langle M_0(t)x, x \rangle &= J^0(t, x; 0) \\
&= \mathbb{E}\left[\langle G \mathring{X}_x^0(T), \mathring{X}_x^0(T) \rangle + \int_t^T \langle Q(s) \mathring{X}_x^0(s), \mathring{X}_x^0(s) \rangle ds \right] \\
&= \mathbb{E}\left[\langle G \Phi(T) \Phi(t)^{-1} x, \Phi(T) \Phi(t)^{-1} x \rangle \right. \\
&\quad\left. + \int_t^T \langle Q(s) \Phi(s) \Phi(t)^{-1} x, \Phi(s) \Phi(t)^{-1} x \rangle ds \right],
\end{aligned}
$$

from which we conclude that M_0 admits the representation (2.2.7). Differentiating $M_0(t)$ shows that M_0 solves Lyapunov equation (2.2.6).   □

Remark 2.2.2 By an argument similar to that used to prove (i)–(ii) of Proposition 2.2.1, we can show that the process ν_t defined in (2.2.3) is given by

$$
\nu_t(s) = B(s)^\top Y_0^0(s) + D(s)^\top Z_0^0(s) + S(s) X_0^0(s) + \rho(s),
$$

where $(X_0^0(s), Y_0^0(s), Z_0^0(s))$ is the adapted solution to the following (decoupled) linear FBSDE on $[t, T]$:

$$
\begin{cases}
dX_0^0(s) = (AX_0^0 + b)ds + (CX_0^0 + \sigma)dW(s), \\
dY_0^0(s) = -(A^\top Y_0^0 + C^\top Z_0^0 + QX_0^0 + q)ds + Z_0^0 dW(s), \\
X_0^0(t) = 0, \quad Y_0^0(T) = GX_0^0(T) + g.
\end{cases} \tag{2.2.10}
$$

We leave the verification of this fact to the interested readers.

The following result is concerned with the convexity of the cost functional, whose proof is straightforward, by making use of (2.2.2) and (2.2.4), respectively.

Proposition 2.2.3 *Let (H1)–(H2) hold, and let $t \in [0, T)$ be given. Then the following statements are equivalent:*

(i) *$u \mapsto J(t, x; u)$ is convex for some $x \in \mathbb{R}^n$ (or for all $x \in \mathbb{R}^n$);*
(ii) *$u \mapsto J^0(t, x; u)$ is convex for some $x \in \mathbb{R}^n$ (or for all $x \in \mathbb{R}^n$);*

(iii) $J^0(t, 0; u) \geq 0$ *for all* $u \in \mathcal{U}[t, T]$;
(iv) $M_2(t)$ *is a nonnegative operator, i.e.,* $M_2(t) \geq 0$.

It is not hard for us to state and prove a similar result for the cases of strict and uniform convexities. We omit the details here. Further, from the representation (2.2.2) and Proposition 1.3.1 in Chap. 1, one has the following result.

Corollary 2.2.4 *Let (H1)–(H2) hold, and let* $(t, x) \in [0, T) \times \mathbb{R}^n$ *be a given initial pair. If Problem* (SLQ) *is finite at* (t, x), *then* $M_2(t) \geq 0$.

The above tells us that the finiteness of Problem (SLQ) (at (t, x)) implies the convexity of $u \mapsto J(t, x; u)$. On the other hand, if $M_2(t) = 0$, although $u \mapsto J(t, x; u)$ is still convex (linear, in fact), but, as long as $M_1(t)x + v_t \neq 0$, Problem (SLQ) will not be finite. Thus, the convexity of $u \mapsto J(t, x; u)$ is not enough to ensure the finiteness of Problem (SLQ). The following example further shows that even the strict convexity of $u \mapsto J(t, x; u)$ is not sufficient for the finiteness of Problem (SLQ).

Example 2.2.5 Consider the one-dimensional controlled SDE

$$\begin{cases} dX(s) = u(s)ds + X(s)dW(s), & s \in [t, T], \\ X(t) = x, \end{cases}$$

and the cost functional

$$J(t, x; u) = \mathbb{E}\left[-X(T)^2 + \int_t^T e^{T-s}u(s)^2 ds \right].$$

We claim that if $T - t = 1$, then

$$J(t, 0; u) > 0, \quad \forall u \in \mathcal{U}[t, T] \setminus \{0\},$$

which, similar to Proposition 2.2.3, is equivalent to the strict convexity of $u \mapsto J(t, x; u)$, but

$$V(t, x) = -\infty, \quad \forall x \neq 0.$$

To verify this claim, let $u \in \mathcal{U}[t, T]$ and X be the corresponding state process with initial state x. By the variation of constants formula,

$$X(s) = xe^{W(s)-W(t)-\frac{1}{2}(s-t)} + \int_t^s e^{W(s)-W(r)-\frac{1}{2}(s-r)}u(r)dr, \quad s \in [t, T].$$

Taking $x = 0$ and noting that $W(T) - W(r)$ is independent of $\mathcal{F}_r$, we have

$$\mathbb{E}\big[X(T)^2\big] = \mathbb{E}\bigg[\int_t^T e^{W(T)-W(r)-\frac{1}{2}(T-r)}u(r)dr\bigg]^2$$

$$\leqslant (T-t)\mathbb{E}\int_t^T e^{2[W(T)-W(r)]-(T-r)}u(r)^2 dr$$

$$= (T-t)\int_t^T \mathbb{E}e^{2[W(T)-W(r)]-(T-r)}\mathbb{E}\big[u(r)^2\big]dr$$

$$= (T-t)\int_t^T e^{T-r}\mathbb{E}\big[u(r)^2\big]dr, \tag{2.2.11}$$

where the inequality follows from Hölder's inequality. According to Hölder's inequality, if the equality holds in the second inequality in (2.2.11), then there exists a constant c such that

$$e^{W(T)-W(r)-\frac{1}{2}(T-r)}u(r) = c, \quad \text{a.e. } r \in [t, T], \text{ a.s.} \tag{2.2.12}$$

Since $W(T) - W(r)$ is independent of $\mathcal{F}_r$ and $u(r)$ is $\mathcal{F}_r$-measurable, by taking conditional expectations with respect to $\mathcal{F}_r$ on both sides of (2.2.12), we obtain

$$u(r) = c, \quad \text{a.e. } r \in [t, T], \text{ a.s.} \tag{2.2.13}$$

Combining (2.2.12)–(2.2.13), we conclude that the equality holds in the second inequality in (2.2.11) if and only if $u = 0$. Thus, if $T - t = 1$ and $u \neq 0$, then

$$J(t, 0; u) = \mathbb{E}\bigg[-X(T)^2 + \int_t^T e^{T-s}u(s)^2 ds\bigg] > 0.$$

On the other hand, if $x \neq 0$ and $u(s) = \lambda e^{W(s)-W(t)-\frac{1}{2}(s-t)}$, $\lambda \in \mathbb{R}$, then

$$X(T) = [x + (T-t)\lambda]e^{W(T)-W(t)-\frac{1}{2}(T-t)},$$

and hence,

$$J(t, x; u) = -[x + (T-t)\lambda]^2 e^{T-t} + \lambda^2(T-t)e^{T-t}.$$

When $T - t = 1$, the above reduces to

$$J(t, x; u) = -\big(2\lambda x + x^2\big)e.$$

Letting $|\lambda| \to \infty$ along $\lambda x > 0$, we get $V(t, x) = -\infty$. This proves our claim.

In order to study the finiteness of Problem (SLQ) at t, let us consider, for $\varepsilon > 0$, the new cost functional $J_\varepsilon(t, x; u)$ defined by

$$J_\varepsilon(t, x; u) \triangleq J(t, x; u) + \varepsilon \mathbb{E} \int_t^T |u(s)|^2 ds, \quad u \in \mathcal{U}[t, T].$$

We denote by Problem (SLQ)$_\varepsilon$ the SLQ problem of minimizing $J_\varepsilon(t, x; u)$ subject to the same state equation (2.1.1). The value function of Problem (SLQ)$_\varepsilon$ is denoted by $V_\varepsilon(t, x)$. Similar to Problem (SLQ)^0, we adopt the notation Problem (SLQ)$^0_\varepsilon$, $J^0_\varepsilon(t, x; u)$, $V^0_\varepsilon(t, x)$ when $b, \sigma, g, q, \rho = 0$.

Proposition 2.2.6 *Let (H1)–(H2) hold.*

(i) *If Problem (SLQ) is finite at (t, x), then*

$$\lim_{\varepsilon \to 0} V_\varepsilon(t, x) = V(t, x). \tag{2.2.14}$$

(ii) *Problem (SLQ)^0 is finite at time t if and only if $M_2(t) \geqslant 0$ and there exists a matrix $P(t) \in \mathbb{S}^n$ such that*

$$V^0(t, x) = \langle P(t)x, x \rangle, \quad \forall x \in \mathbb{R}^n. \tag{2.2.15}$$

Proof (i) It follows from Proposition 1.3.2 of Chap. 1.

(ii) The sufficiency is trivial. Let us prove the necessity. First of all, by Corollary 2.2.4, we have $M_2(t) \geqslant 0$. So for each $\varepsilon > 0$,

$$\begin{aligned}
J^0_\varepsilon(t, x; u) &= J^0(t, x; u) + \varepsilon \mathbb{E} \int_t^T |u(s)|^2 ds \\
&= \langle [M_2(t) + \varepsilon I]u, u \rangle + 2\langle M_1(t)x, u \rangle + \langle M_0(t)x, x \rangle \\
&= \|[M_2(t) + \varepsilon I]^{\frac{1}{2}} u + [M_2(t) + \varepsilon I]^{-\frac{1}{2}} M_1(t)x\|^2 \\
&\quad + \langle \{M_0(t) - M_1(t)^*[M_2(t) + \varepsilon I]^{-1} M_1(t)\}x, x \rangle,
\end{aligned}$$

where $M_1(t)^*$ is the adjoint operator of $M_1(t)$. Clearly, with the notation

$$P_\varepsilon(t) \equiv M_0(t) - M_1(t)^*[M_2(t) + \varepsilon I]^{-1} M_1(t) \in \mathbb{S}^n,$$

one has

$$V^0_\varepsilon(t, x) = \inf_{u \in \mathcal{U}[t,T]} J^0_\varepsilon(t, x; u) = \langle P_\varepsilon(t)x, x \rangle.$$

On the other hand, we have by (i) that

$$\lim_{\varepsilon \to 0} \langle P_\varepsilon(t)x, x \rangle = \lim_{\varepsilon \to 0} V^0_\varepsilon(t, x) = V^0(t, x).$$

Since Problem (SLQ)^0 is finite at t, the above holds for all $x \in \mathbb{R}^n$. Therefore, the limit $P(t) \equiv \lim_{\varepsilon \to 0} P_\varepsilon(t)$ exists and satisfies (2.2.15). □

2.3   Open-Loop Solvability and FBSDEs

The aim of this section is to provide a characterization of the open-loop solvability of Problem (SLQ) in terms of FBSDEs. We begin with a simple property of the cost functional, which is a consequence of Proposition 2.2.1 and Remark 2.2.2.

Proposition 2.3.1 *Let (H1)–(H2) hold, and let $t \in [0, T)$ be given. For any $x \in \mathbb{R}^n$, $\lambda \in \mathbb{R}$, and $u, v \in \mathcal{U}[t, T]$, we have*

$$J(t, x; u + \lambda v) = J(t, x; u) + \lambda^2 J^0(t, 0; v)$$
$$+ 2\lambda \mathbb{E} \int_t^T \big(B^\top Y + D^\top Z + SX + Ru + \rho, v\big)ds,$$

where (X, Y, Z) is the adapted solution to the (decoupled) FBSDE on $[t, T]$:

$$\begin{cases} dX(s) = (AX + Bu + b)ds + (CX + Du + \sigma)dW, \\ dY(s) = -\big(A^\top Y + C^\top Z + QX + S^\top u + q\big)ds + ZdW, \\ X(t) = x, \quad Y(T) = GX(T) + g. \end{cases} \qquad (2.3.1)$$

Consequently, the mapping $u \mapsto J(t, x; u)$ is Fréchet differentiable, and its Fréchet derivative at u is given by

$$\mathcal{D}_u J(t, x; u)(s) = 2\big[B(s)^\top Y(s) + D(s)^\top Z(s) + S(s)X(s)$$
$$+ R(s)u(s) + \rho(s)\big], \quad s \in [t, T]. \qquad (2.3.2)$$

Proof By the representations (2.2.2) and (2.2.4) of the cost functional, we have

$$\begin{aligned} J(t, x; u + \lambda v) &= \langle M_2(t)(u + \lambda v), u + \lambda v \rangle + 2\langle M_1(t)x, u + \lambda v \rangle \\ &\quad + \langle M_0(t)x, x \rangle + 2\langle u + \lambda v, \nu_t \rangle + 2\langle x, y_t \rangle + c_t \\ &= J(t, x; u) + \lambda^2 J^0(t, 0; v) + 2\lambda \langle M_2(t)u + M_1(t)x + \nu_t, v \rangle. \end{aligned}$$

According to Proposition 2.2.1 and Remark 2.2.2,

$$M_2(t)u + M_1(t)x + \nu_t = B^\top (\mathring{Y}_0^u + \mathring{Y}_x^0 + Y_0^0) + D^\top (\mathring{Z}_0^u + \mathring{Z}_x^0 + Z_0^0)$$
$$+ S(\mathring{X}_0^u + \mathring{X}_x^0 + X_0^0) + Ru + \rho.$$

Set $X = \mathring{X}_0^u + \mathring{X}_x^0 + X_0^0$, $Y = \mathring{Y}_0^u + \mathring{Y}_x^0 + Y_0^0$, and $Z = \mathring{Z}_0^u + \mathring{Z}_x^0 + Z_0^0$. Then by the linearity of the FBSDEs (2.2.5) and (2.2.10), we see that (X, Y, Z) solves (2.3.1). The rest of the proof is now clear. $\qquad \square$

Theorem 2.3.2 *Let (H1)–(H2) hold, and let the initial pair $(t, x) \in [0, T) \times \mathbb{R}^n$ be given. A control $u \in \mathcal{U}[t, T]$ is open-loop optimal for (t, x) if and only if*

     (i)   *the mapping $u \mapsto J^0(t, 0; u)$ is convex,*

(ii) *the adapted solution (X, Y, Z) to the (decoupled) FBSDE* (2.3.1) *satisfies the stationarity condition*

$$B^\top Y + D^\top Z + SX + Ru + \rho = 0, \quad \text{a.e. } s \in [t, T], \text{ a.s.}$$

Proof Let $u \in \mathcal{U}[t, T]$ and (X, Y, Z) be the adapted solution to FBSDE (2.3.1). We first observe that u is open-loop optimal for (t, x) if and only if

$$J(t, x; u + \lambda v) - J(t, x; u) \geqslant 0, \quad \forall \lambda \in \mathbb{R}, \ \forall v \in \mathcal{U}[t, T].$$

According to Proposition 2.3.1, for any $\lambda \in \mathbb{R}$ and $v \in \mathcal{U}[t, T]$,

$$\begin{aligned}
& J(t, x; u + \lambda v) - J(t, x; u) \\
& = \lambda^2 J^0(t, 0; v) + \lambda \mathbb{E} \int_t^T \langle \mathcal{D}_u J(t, x; u)(s), v(s) \rangle ds,
\end{aligned}$$

where $\mathcal{D}_u J(t, x; u)$ is given by (2.3.2). Thus, (i) and (ii) trivially imply that u is open-loop optimal. Conversely, if u is open-loop optimal for (t, x), then for fixed but arbitrary $v \in \mathcal{U}[t, T]$,

$$\lambda^2 J^0(t, 0; v) + \lambda \mathbb{E} \int_t^T \langle \mathcal{D}_u J(t, x; u)(s), v(s) \rangle ds$$

is a nonnegative and quadratic function of λ. Thus, we must have

$$J^0(t, 0; v) \geqslant 0, \quad \mathbb{E} \int_t^T \langle \mathcal{D}_u J(t, x; u)(s), v(s) \rangle ds = 0.$$

Since v is arbitrary, the necessity follows. $\square$

From the above result, we see that if Problem (SLQ) admits an open-loop optimal control u for the initial pair (t, x), then u is determined by the following system of equations:

$$\begin{cases}
dX = (AX + Bu + b)ds + (CX + Du + \sigma)dW, & s \in [t, T], \\
dY = -(A^\top Y + C^\top Z + QX + S^\top u + q)ds + ZdW, & s \in [t, T], \\
X(t) = x, \quad Y(T) = GX(T) + g, \\
B^\top Y + D^\top Z + SX + Ru + \rho = 0, \quad \text{a.e. } s \in [t, T], \text{ a.s.}
\end{cases} \tag{2.3.3}$$

We call (2.3.3) the *optimality system* of Problem (SLQ). Note that the stationarity condition (the last equation in (2.3.3)) brings a coupling into the FBSDE in (2.3.3). Thus, to find an open-loop optimal control of Problem (SLQ), one actually needs to solve a *coupled* FBSDE.

2.4   Closed-Loop Solvability and Riccati Equation

In this section we focus on the closed-loop solvability of Problem (SLQ). In order to obtain an analytical characterization of the closed-loop optimal strategies, we introduce the following nonlinear ordinary differential equation:

$$
\begin{cases}
\dot{P}(s) + P(s)A(s) + A(s)^\top P(s) + C(s)^\top P(s)C(s) + Q(s) \\
\quad - \big[P(s)B(s) + C(s)^\top P(s)D(s) + S(s)^\top\big]\big[R(s) + D(s)^\top P(s)D(s)\big]^\dagger \\
\quad \times \big[B(s)^\top P(s) + D(s)^\top P(s)C(s) + S(s)\big] = 0, \\
P(T) = G,
\end{cases}
\tag{2.4.1}
$$

where $M^\dagger$ denotes the Moore-Penrose pseudoinverse of a matrix M. Equation (2.4.1) is called the *Riccati equation* associated with Problem (SLQ). Notice that the Eq. (2.4.1) is symmetric. By a solution to (2.4.1) we mean a continuous $\mathbb{S}^n$-valued function that satisfies (2.4.1) for almost all s. Using the notation

$$
\begin{cases}
\mathcal{Q}(s, P) = PA(s) + A(s)^\top P + C(s)^\top PC(s) + Q(s), \\
\mathcal{S}(s, P) = B(s)^\top P + D(s)^\top PC(s) + S(s), \\
\mathcal{R}(s, P) = R(s) + D(s)^\top PD(s),
\end{cases}
\tag{2.4.2}
$$

we can rewrite the Riccati equation (2.4.1) as

$$
\begin{cases}
\dot{P}(s) + \mathcal{Q}(s, P(s)) - \mathcal{S}(s, P(s))^\top \mathcal{R}(s, P(s))^\dagger \mathcal{S}(s, P(s)) = 0, \\
P(T) = G.
\end{cases}
$$

When $P(s)$ is a solution of (2.4.1), to simplify notation we will frequently suppress the variable s and write $\mathcal{Q}(s, P(s)), \mathcal{S}(s, P(s))$, and $\mathcal{R}(s, P(s))$ as $\mathcal{Q}(P), \mathcal{S}(P)$, and $\mathcal{R}(P)$, respectively.

Definition 2.4.1 Let $P \in C([t, T]; \mathbb{S}^n)$ be a solution to the Riccati equation (2.4.1) on $[t, T]$. It is called *regular* if

 (i) $\mathcal{R}(s, P(s)) \geqslant 0$ for a.e. $s \in [t, T]$,
 (ii) $\mathscr{R}(\mathcal{S}(s, P(s))) \subseteq \mathscr{R}(\mathcal{R}(s, P(s)))$ for a.e. $s \in [t, T]$,
 (iii) $\mathcal{R}(P)^\dagger \mathcal{S}(P) \in \Theta[t, T] \equiv L^2(t, T; \mathbb{R}^{m \times n})$,

where $\mathscr{R}(M)$ denotes the range of a matrix M. The Riccati equation (2.4.1) is said to be *regularly solvable* if it admits a regular solution.

Remark 2.4.2 Under the assumptions (H1)–(H2), if $P \in C([t, T]; \mathbb{S}^n)$ is a regular solution to the Riccati equation (2.4.1) on $[t, T]$, then the square-integrability of

$$
\Theta \equiv -(R + D^\top PD)^\dagger(B^\top P + D^\top PC + S) = -\mathcal{R}(P)^\dagger \mathcal{S}(P)
$$

guarantees the existence of a unique adapted solution (η, ζ) to the BSDE

$$
\begin{cases}
d\eta(s) = -\big\{[A(s) + B(s)\Theta(s)]^\top \eta(s) + [C(s) + D(s)\Theta(s)]^\top \zeta(s) \\
\qquad\qquad + [C(s) + D(s)\Theta(s)]^\top P(s)\sigma(s) + \Theta(s)^\top \rho(s) \\
\qquad\qquad + P(s)b(s) + q(s)\big\}ds + \zeta(s)dW(s), \quad s \in [t, T], \\
\eta(T) = g.
\end{cases}
\tag{2.4.3}
$$

Now we present the main result of this section, which establishes the equivalence between the closed-loop solvability of Problem (SLQ) and the regular solvability of the associated Riccati equation (2.4.1).

Theorem 2.4.3 *Let (H1)–(H2) hold. Then Problem (SLQ) is closed-loop solvable on $[t, T]$ if and only if the following two conditions hold:*

(i) The Riccati equation (2.4.1) admits a regular solution $P \in C([t, T]; \mathbb{S}^n)$.
(ii) Let (η, ζ) be the adapted solution to the BSDE (2.4.3) and define

$$
\kappa(s) \triangleq B(s)^\top \eta(s) + D(s)^\top \zeta(s) + D(s)^\top P(s)\sigma(s) + \rho(s),
$$
$$
v(s) \triangleq -\mathcal{R}(s, P(s))^\dagger \kappa(s).
$$

Then

$$
\kappa(s) \in \mathcal{R}(\mathcal{R}(s, P(s))), \quad \text{a.e. a.s.} \tag{2.4.4}
$$
$$
v \in \mathcal{U}[t, T] \equiv L_\mathbb{F}^2(t, T; \mathbb{R}^m). \tag{2.4.5}
$$

In this case, the closed-loop optimal strategy $(\bar\Theta, \bar v)$ admits the following representation:

$$
\bar\Theta = \Theta + [I - \mathcal{R}(P)^\dagger \mathcal{R}(P)]\Pi, \quad \bar v = v + [I - \mathcal{R}(P)^\dagger \mathcal{R}(P)]\pi, \tag{2.4.6}
$$

with $(\Pi, \pi) \in \Theta[t, T] \times \mathcal{U}[t, T]$ being arbitrary. Further, the value function is given by

$$
V(t, x) = \mathbb{E}\Big\{\langle P(t)x, x\rangle + 2\langle \eta(t), x\rangle + \int_t^T \big[\langle P(s)\sigma(s), \sigma(s)\rangle + 2\langle \eta(s), b(s)\rangle
$$
$$
+ 2\langle \zeta(s), \sigma(s)\rangle - \langle \mathcal{R}(s, P(s))^\dagger \kappa(s), \kappa(s)\rangle\big]ds\Big\}.
$$

To prove the above result we need the following proposition which is a consequence of Theorem 2.3.2.

Proposition 2.4.4 *Let (H1)–(H2) hold. Let $(\bar{\Theta}, \bar{v}) \in \Theta[t, T] \times \mathcal{U}[t, T]$ be a closed-loop optimal strategy of Problem* (SLQ) *on* $[t, T]$. *Then the adapted solution* $(\mathbb{X}, \mathbb{Y}, \mathbb{Z})$ *to the matrix FBSDE*

$$\begin{cases} d\mathbb{X}(s) = \big(A + B\bar{\Theta}\big)\mathbb{X}ds + \big(C + D\bar{\Theta}\big)\mathbb{X}dW(s), \quad s \in [t, T], \\ d\mathbb{Y}(s) = -\big[A^\top\mathbb{Y} + C^\top\mathbb{Z} + \big(Q + S^\top\bar{\Theta}\big)\mathbb{X}\big]ds + \mathbb{Z}dW(s), \quad s \in [t, T], \\ \mathbb{X}(t) = I, \quad \mathbb{Y}(T) = G\mathbb{X}(T), \end{cases}$$

satisfies the following condition:

$$B(s)^\top\mathbb{Y}(s) + D(s)^\top\mathbb{Z}(s) + \big[S(s) + R(s)\bar{\Theta}(s)\big]\mathbb{X}(s) = 0,$$
$$\text{a.e. } s \in [t, T], \text{ a.s.} \tag{2.4.7}$$

Proof Consider the state equation

$$\begin{cases} dX(s) = [(A + B\bar{\Theta})X + Bv + b]ds \\ \qquad\qquad + [(C + D\bar{\Theta})X + Dv + \sigma]dW(s), \quad s \in [t, T], \\ X(t) = x, \end{cases}$$

and the cost functional

$$\tilde{J}(t, x; v) \triangleq J(t, x; \bar{\Theta}X + v).$$

A straightforward calculation shows that

$$\tilde{J}(t, x; v) = \mathbb{E}\Big\{ \langle GX(T), X(T) \rangle + 2\langle g, X(T) \rangle$$
$$+ \int_t^T \Big[\Big\langle \begin{pmatrix} \tilde{Q} & \tilde{S}^\top \\ \tilde{S} & R \end{pmatrix}\begin{pmatrix} X \\ v \end{pmatrix}, \begin{pmatrix} X \\ v \end{pmatrix}\Big\rangle + 2\Big\langle \begin{pmatrix} \tilde{q} \\ \rho \end{pmatrix}, \begin{pmatrix} X \\ v \end{pmatrix}\Big\rangle\Big]ds\Big\},$$

where

$$\tilde{Q} = Q + \bar{\Theta}^\top S + S^\top\bar{\Theta} + \bar{\Theta}^\top R\bar{\Theta}, \quad \tilde{S} = S + R\bar{\Theta}, \quad \tilde{q} = q + \bar{\Theta}^\top\rho.$$

According to Proposition 2.1.5, $\bar{v}$ is an open-loop optimal control for the above LQ problem for any initial state x. Thus, by Theorem 2.3.2, we have

$$B^\top\bar{Y} + D^\top\bar{Z} + (S + R\bar{\Theta})\bar{X} + R\bar{v} + \rho = 0, \quad \text{a.e. } s \in [t, T] \text{ a.s.} \tag{2.4.8}$$

where $(\bar{Y}, \bar{Z})$ is the adapted solution to the following BSDE:

$$\begin{cases} d\bar{Y}(s) = -\big[(A + B\bar{\Theta})^\top\bar{Y} + (C + D\bar{\Theta})^\top\bar{Z} + \tilde{Q}\bar{X} + \tilde{S}^\top\bar{v} + \tilde{q}\big]ds + \bar{Z}d W, \\ \bar{Y}(T) = G\bar{X}(T) + g. \end{cases}$$

By (2.4.8), the SDE in the above can be rewritten as

$$
\begin{aligned}
d\bar{Y} &= -\big[(A + B\bar{\Theta})^{\top}\bar{Y} + (C + D\bar{\Theta})^{\top}\bar{Z} + (Q + \bar{\Theta}^{\top}S + S^{\top}\bar{\Theta} + \bar{\Theta}^{\top}R\bar{\Theta})\bar{X} \\
&\quad + (S + R\bar{\Theta})^{\top}\bar{v} + q + \bar{\Theta}^{\top}\rho\big]ds + \bar{Z}dW \\
&= -\big\{A^{\top}\bar{Y} + C^{\top}\bar{Z} + (Q + S^{\top}\bar{\Theta})\bar{X} + S^{\top}\bar{v} + q \\
&\quad + \bar{\Theta}^{\top}[B^{\top}\bar{Y} + D^{\top}\bar{Z} + (S + R\bar{\Theta})\bar{X} + R\bar{v} + \rho]\big\}ds + \bar{Z}dW \\
&= -\big[A^{\top}\bar{Y} + C^{\top}\bar{Z} + (Q + S^{\top}\bar{\Theta})\bar{X} + S^{\top}\bar{v} + q\big]ds + \bar{Z}dW.
\end{aligned}
$$

Thus, for any initial state x, the adapted solution $(\bar{X}, \bar{Y}, \bar{Z})$ to the FBSDE

$$
\begin{cases}
d\bar{X} = \big[(A + B\bar{\Theta})\bar{X} + B\bar{v} + b\big]ds + \big[(C + D\bar{\Theta})\bar{X} + D\bar{v} + \sigma\big]dW, \\
d\bar{Y} = -\big[A^{\top}\bar{Y} + C^{\top}\bar{Z} + (Q + S^{\top}\bar{\Theta})\bar{X} + S^{\top}\bar{v} + q\big]ds + \bar{Z}dW, \\
\bar{X}(t) = x, \quad \bar{Y}(T) = G\bar{X}(T) + g,
\end{cases}
\tag{2.4.9}
$$

satisfies (2.4.8). Since x is arbitrary and $(\bar{\Theta}, \bar{v})$ is independent of x, by subtracting solutions corresponding to x and 0, the latter from the former, we see that for any $x \in \mathbb{R}^n$,

$$
B^{\top}Y + C^{\top}Z + (S + R\bar{\Theta})X = 0, \quad \text{a.e. } s \in [t, T], \text{ a.s.}
$$

where (X, Y, Z) is the adapted solution to the following FBSDE:

$$
\begin{cases}
dX(s) = (A + B\bar{\Theta})X ds + (C + D\bar{\Theta})X dW(s), \quad s \in [t, T], \\
dY(s) = -\big[A^{\top}Y + C^{\top}Z + (Q + S^{\top}\bar{\Theta})X\big]ds + ZdW(s), \quad s \in [t, T], \\
X(t) = x, \quad Y(T) = GX(T).
\end{cases}
$$

The conclusion follows now from the fact that $X(s) = \mathbb{X}(s)x$, $Y(s) = \mathbb{Y}(s)x$, and $Z(s) = \mathbb{Z}(s)x$. $\qquad\square$

Proof of Theorem 2.4.3. We begin with the necessity. Suppose that $(\bar{\Theta}, \bar{v})$ is a closed-loop optimal strategy of Problem (SLQ) over $[t, T]$. Then, according to Proposition 2.4.4, (2.4.7) holds. Notice that $\mathbb{X}^{-1}$ exists and satisfies the following SDE:

$$
\begin{cases}
d(\mathbb{X}^{-1}) = \mathbb{X}^{-1}\big[(C + D\bar{\Theta})^2 - (A + B\bar{\Theta})\big]ds - \mathbb{X}^{-1}(C + D\bar{\Theta})dW, \quad s \in [t, T], \\
\mathbb{X}(t)^{-1} = I_n.
\end{cases}
$$

Thus we can define

$$
P(s) = \mathbb{Y}(s)\mathbb{X}(s)^{-1}, \quad \Gamma(s) = \mathbb{Z}(s)\mathbb{X}(s)^{-1}; \quad s \in [t, T].
$$

Post-multiplying (2.4.7) by $\mathbb{X}(s)^{-1}$, we obtain

$$
B^{\top}P + D^{\top}\Gamma + S + R\bar{\Theta} = 0, \quad \text{a.e. } s \in [t, T], \text{ a.s.}
\tag{2.4.10}
$$

and by Itô's formula, we have

$$
\begin{aligned}
dP &= -\big[A^\top \mathbb{Y} + C^\top \mathbb{Z} + (Q + S^\top \bar{\Theta})\mathbb{X}\big]\mathbb{X}^{-1}ds + \mathbb{Z}\mathbb{X}^{-1}dW \\
&\quad + \mathbb{Y}\mathbb{X}^{-1}\big[(C + D\bar{\Theta})^2 - (A + B\bar{\Theta})\big]ds - \mathbb{Y}\mathbb{X}^{-1}(C + D\bar{\Theta})dW \\
&\quad - \mathbb{Z}\mathbb{X}^{-1}(C + D\bar{\Theta})ds \\
&= \big\{ - A^\top P - C^\top \Gamma - Q - S^\top \bar{\Theta} + P\big[(C + D\bar{\Theta})^2 - (A + B\bar{\Theta})\big] \\
&\quad - \Gamma(C + D\bar{\Theta})\big\}ds + \big[\Gamma - P(C + D\bar{\Theta})\big]dW.
\end{aligned}
$$

Set $\Lambda = \Gamma - P(C + D\bar{\Theta})$. Then the above reduces to

$$
dP = \big[- \mathcal{Q}(P) - \Lambda C - C^\top \Lambda - \Lambda D\bar{\Theta} - \mathcal{S}(P)^\top \bar{\Theta} \big]ds + \Lambda dW.
$$

Note also that $P(T) = G$. Thus, (P, Λ) is the adapted solution to a BSDE with deterministic coefficients. Consequently, P must be deterministic and Λ must be zero, which implies $\Gamma = P(C + D\bar{\Theta})$ and

$$
\dot{P} + \mathcal{Q}(P) + \mathcal{S}(P)^\top \bar{\Theta} = 0. \tag{2.4.11}
$$

Substituting $\Gamma = P(C + D\bar{\Theta})$ into (2.4.10), we obtain

$$
\mathcal{S}(P) + \mathcal{R}(P)\bar{\Theta} = 0, \quad \text{a.e. } s \in [t, T], \tag{2.4.12}
$$

which, by Proposition A.1.5 in Appendix, implies that

$$
\mathscr{R}(\mathcal{S}(P)) \subseteq \mathscr{R}(\mathcal{R}(P)), \text{ a.e.}, \qquad \Theta \equiv -\mathcal{R}(P)^\dagger \mathcal{S}(P) \in \boldsymbol{\Theta}[t, T],
$$

and that

$$
\bar{\Theta} = \Theta + [I - \mathcal{R}(P)^\dagger \mathcal{R}(P)]\Pi,
$$

for some $\Pi \in \boldsymbol{\Theta}[t, T]$. Using (2.4.12), one can rewrite (2.4.11) as

$$
\dot{P} + \mathcal{Q}(P) + \bar{\Theta}^\top \mathcal{R}(P)^\top \bar{\Theta} = 0.
$$

Since $P(T) = G \in \mathbb{S}^n$ and Q, R are symmetric, by the uniqueness of solutions, we must have $P \in C([t, T]; \mathbb{S}^n)$. We see now that $(R + D^\top PD)$ is symmetric. Further, making use of (2.4.12), we have

$$
\begin{aligned}
\mathcal{S}(P)^\top \bar{\Theta} &= \mathcal{S}(P)^\top \Theta + \mathcal{S}(P)^\top [I - \mathcal{R}(P)^\dagger \mathcal{R}(P)]\Pi \\
&= -\mathcal{S}(P)^\top \mathcal{R}(P)^\dagger \mathcal{S}(P) - \bar{\Theta}^\top \mathcal{R}(P)[I - \mathcal{R}(P)^\dagger \mathcal{R}(P)]\Pi \\
&= -\mathcal{S}(P)^\top \mathcal{R}(P)^\dagger \mathcal{S}(P).
\end{aligned}
$$

Plugging the above into (2.4.11), we obtain the Riccati equation (2.4.1).

To determine $\bar{v}$, let $(\bar{X}, \bar{Y}, \bar{Z})$ be the adapted solution to the FBSDE (2.4.9) and define for $s \in [t, T]$,

$$\eta(s) = \bar{Y}(s) - P(s)\bar{X}(s),$$
$$\zeta(s) = \bar{Z}(s) - P(s)\{[C(s) + D(s)\bar{\Theta}(s)]\bar{X}(s) + D(s)\bar{v}(s) + \sigma(s)\}.$$

Then, by making use of Itô's formula and (2.4.11), we have

$$\begin{aligned}
d\eta &= d\bar{Y} - \dot{P}\bar{X}ds - Pd\bar{X} \\
&= -\big[A^{\top}\bar{Y} + C^{\top}\bar{Z} + (Q + S^{\top}\bar{\Theta})\bar{X} + S^{\top}\bar{v} + q\big]ds + \bar{Z}d W - \dot{P}\bar{X}ds \\
&\quad - P\big[(A + B\bar{\Theta})\bar{X} + B\bar{v} + b\big]ds - P\big[(C + D\bar{\Theta})\bar{X} + D\bar{v} + \sigma\big]dW \\
&= -\Big[A^{\top}(\eta + P\bar{X}) + C^{\top}\{\zeta + P[(C + D\bar{\Theta})\bar{X} + D\bar{v} + \sigma]\} + (Q + S^{\top}\bar{\Theta})\bar{X} \\
&\quad + S^{\top}\bar{v} + q + \dot{P}\bar{X} + P[(A + B\bar{\Theta})\bar{X} + B\bar{v} + b]\Big]ds + \zeta dW \\
&= -\big[A^{\top}\eta + C^{\top}\zeta + \mathcal{S}(P)^{\top}\bar{v} + C^{\top}P\sigma + Pb + q\big]ds + \zeta dW.
\end{aligned}$$

Recall from the proof of Proposition 2.4.4 that $(\bar{X}, \bar{Y}, \bar{Z})$ satisfies (2.4.8). Substituting for $\bar{Y}$ and $\bar{Z}$ in Eq. (2.4.8) and making use of (2.4.12), we obtain

$$\begin{aligned}
0 &= B^{\top}(\eta + P\bar{X}) + D^{\top}\big[\zeta + P(C + D\bar{\Theta})\bar{X} + PD\bar{v} + P\sigma\big] \\
&\quad + (S + R\bar{\Theta})\bar{X} + R\bar{v} + \rho \\
&= B^{\top}\eta + D^{\top}\zeta + D^{\top}P\sigma + \rho + \mathcal{R}(P)\bar{v}.
\end{aligned}$$

Now we can repeat the argument employed in the proof of Proposition A.1.5 in Appendix, replacing $L^2(\mathcal{I}; \mathbb{R}^{m \times k})$ by $\mathcal{U}[t, T] \equiv L^2_{\mathbb{F}}(t, T; \mathbb{R}^m)$, to obtain

$$\kappa \triangleq B^{\top}\eta + D^{\top}\zeta + D^{\top}P\sigma + \rho \in \mathcal{R}(\mathcal{R}(P)), \quad \text{a.e. a.s.}$$
$$v \triangleq -\mathcal{R}(P)^{\dagger}\kappa \in \mathcal{U}[t, T],$$

and that

$$\bar{v} = v + [I - \mathcal{R}(P)^{\dagger}\mathcal{R}(P)]\pi,$$

for some $\pi \in \mathcal{U}[t, T]$. Consequently, by (2.4.12),

$$\begin{aligned}
\mathcal{S}(P)^{\top}\bar{v} &= -\mathcal{S}(P)^{\top}\mathcal{R}(P)^{\dagger}\kappa + \mathcal{S}(P)^{\top}[I - \mathcal{R}(P)^{\dagger}\mathcal{R}(P)]\pi \\
&= \Theta^{\top}(B^{\top}\eta + D^{\top}\zeta + D^{\top}P\sigma + \rho) - \bar{\Theta}^{\top}\mathcal{R}(P)[I - \mathcal{R}(P)^{\dagger}\mathcal{R}(P)]\pi \\
&= \Theta^{\top}(B^{\top}\eta + D^{\top}\zeta + D^{\top}P\sigma + \rho),
\end{aligned}$$

and hence

$$A^\top \eta + C^\top \zeta + \mathcal{S}(P)^\top \bar{v} + C^\top P\sigma + Pb + q$$
$$= (A + B\Theta)^\top \eta + (C + D\Theta)^\top \zeta + (C + D\Theta)^\top P\sigma + \Theta^\top \rho + Pb + q.$$

Therefore, (η, ζ) is the adapted solution to the BSDE (2.4.3).

To prove (i) of Definition 2.4.1, as well as the sufficiency, we take arbitrary $x \in \mathbb{R}^n$ and $u \in \mathcal{U}[t, T]$, and let $X(\cdot) \equiv X(\cdot\,; t, x, u)$ be the corresponding state process. Applying Itô's formula to $s \mapsto \langle P(s)X(s), X(s) \rangle$ yields

$$\mathbb{E}\langle GX(T), X(T) \rangle - \langle P(t)x, x \rangle$$
$$= \mathbb{E}\int_t^T \Big[\langle (\dot{P} + \mathcal{Q}(P) - Q)X, X \rangle + 2\langle (B^\top P + D^\top PC)X, u \rangle$$
$$+ \langle D^\top PDu, u \rangle + 2\langle C^\top P\sigma + Pb, X \rangle + 2\langle D^\top P\sigma, u \rangle + \langle P\sigma, \sigma \rangle \Big] ds,$$

and applying Itô's formula to $s \mapsto \langle \eta(s), X(s) \rangle$ yields

$$\mathbb{E}\langle g, X(T) \rangle - \mathbb{E}\langle \eta(t), x \rangle = \mathbb{E}\int_t^T \Big[-\langle \Theta^\top (B^\top \eta + D^\top \zeta + D^\top P\sigma + \rho), X \rangle$$
$$- \langle C^\top P\sigma + Pb + q, X \rangle + \langle B^\top \eta + D^\top \zeta, u \rangle + \langle \eta, b \rangle + \langle \zeta, \sigma \rangle \Big] ds.$$

Substituting for $\mathbb{E}\langle GX(T), X(T) \rangle$ and $\mathbb{E}\langle g, X(T) \rangle$ in the cost functional gives

$$J(t, x; u) = \mathbb{E}\Big\{ \langle P(t)x, x \rangle + 2\langle \eta(t), x \rangle + \int_t^T \Big[\langle P\sigma, \sigma \rangle + 2\langle \eta, b \rangle$$
$$+ 2\langle \zeta, \sigma \rangle + \langle (\dot{P} + \mathcal{Q}(P))X, X \rangle + \langle \mathcal{R}(P)u, u \rangle$$
$$+ 2\langle \mathcal{S}(P)X + B^\top \eta + D^\top \zeta + D^\top P\sigma + \rho, u \rangle$$
$$- 2\langle \Theta^\top (B^\top \eta + D^\top \zeta + D^\top P\sigma + \rho), X \rangle \Big] ds \Big\}. \qquad (2.4.13)$$

Let $\bar{\Theta}$ and $\bar{v}$ be defined in (2.4.6). It is easy to verify that

$$\mathcal{S}(P) = -\mathcal{R}(P)\bar{\Theta}, \quad \dot{P} + \mathcal{Q}(P) = \bar{\Theta}^\top \mathcal{R}(P)\bar{\Theta},$$
$$B^\top \eta + D^\top \zeta + D^\top P\sigma + \rho = -\mathcal{R}(P)\bar{v},$$
$$\Theta^\top (B^\top \eta + D^\top \zeta + D^\top P\sigma + \rho) = -\bar{\Theta}^\top \mathcal{R}(P)\bar{v}.$$

Substituting these equations into (2.4.13) yields

$$J(t, x; u) = \mathbb{E}\Big\{\langle P(t)x, x\rangle + 2\langle\eta(t), x\rangle + \int_t^T \Big[\langle P\sigma, \sigma\rangle + 2\langle\eta, b\rangle + 2\langle\zeta, \sigma\rangle$$
$$+ \langle\bar\Theta^\top\mathcal{R}(P)\bar\Theta X, X\rangle - 2\langle\mathcal{R}(P)(\bar\Theta X + \bar v), u\rangle$$
$$+ \langle\mathcal{R}(P)u, u\rangle + 2\langle\bar\Theta^\top\mathcal{R}(P)\bar v, X\rangle\Big]ds\Big\}.$$

It follows by completing the square that

$$J(t, x; u) = \mathbb{E}\Big\{\langle P(t)x, x\rangle + 2\langle\eta(t), x\rangle + \int_t^T \Big[\langle P\sigma, \sigma\rangle + 2\langle\eta, b\rangle + 2\langle\zeta, \sigma\rangle$$
$$- \langle\mathcal{R}(P)\bar v, \bar v\rangle + \langle\mathcal{R}(P)(u - \bar\Theta X - \bar v), u - \bar\Theta X - \bar v\rangle\Big]ds\Big\}$$
$$= J(t, x; \bar\Theta\bar X + \bar v) + \mathbb{E}\int_t^T \langle\mathcal{R}(P)(u - \bar\Theta X - \bar v), u - \bar\Theta X - \bar v\rangle ds.$$

For any $v \in \mathcal{U}[t, T]$, let $u = \bar\Theta X + v$ with X being the solution to the state equation under the closed-loop strategy $(\bar\Theta, v)$. Then the above implies that

$$J(t, x; \bar\Theta X + v) = J(t, x; \bar\Theta\bar X + \bar v) + \mathbb{E}\int_t^T \langle\mathcal{R}(P)(v - \bar v), v - \bar v\rangle ds.$$

Therefore, $(\bar\Theta, \bar v)$ is a closed-loop optimal strategy if and only if

$$\mathbb{E}\int_t^T \langle\mathcal{R}(P)(v - \bar v), v - \bar v\rangle ds \geqslant 0, \quad \forall v \in \mathcal{U}[t, T],$$

or equivalently,
$$\mathcal{R}(s, P(s)) \geqslant 0, \quad \text{a.e. } s \in [t, T].$$

Finally, the representation of the value function follows from the identity

$$\langle\mathcal{R}(P)\bar v, \bar v\rangle = \langle\mathcal{R}(P)^\dagger\kappa, \kappa\rangle.$$

The proof is complete. □

We point out here that the solution to the Riccati equation (2.4.1) may be non-unique. However, if a regular solution exists, it must be unique.

Corollary 2.4.5 *Let (H1)–(H2) hold. Then the Riccati equation (2.4.1) admits at most one regular solution.*

Proof Consider Problem (SLQ)^0, for which $b, \sigma, g, q, \rho = 0$. Then for any regular solution P, the adapted solution (η, ζ) to the BSDE (2.4.3) is identically zero. Suppose now that P_1 and P_2 are two regular solutions of the Riccati equation (2.4.1). According to the representation of the value function, we have

$$\langle P_1(t)x, x \rangle = V^0(t, x) = \langle P_2(t)x, x \rangle, \quad \forall x \in \mathbb{R}^n,$$

which implies $P_1(t) = P_2(t)$. The identity $P_1(s) = P_2(s)$ for $s \in (t, T)$ can be obtained similarly, by considering Problem $(SLQ)^0$ on $[s, T]$.   □

We conclude this section with an equivalent statement of Theorem 2.4.3, which is even useful for determining whether a closed-loop strategy is optimal.

Theorem 2.4.6 *Let (H1)–(H2) hold. Then a closed-loop strategy* $(\bar{\Theta}, \bar{v}) \in \Theta[t, T] \times \mathcal{U}[t, T]$ *is optimal if and only if*

(i) the solution $P \in C([t, T]; \mathbb{S}^n)$ *to the symmetric Lyapunov type equation*

$$\begin{cases} \dot{P} + \mathcal{Q}(P) + \bar{\Theta}^\top \mathcal{R}(P)\bar{\Theta} + \mathcal{S}(P)^\top \bar{\Theta} + \bar{\Theta}^\top \mathcal{S}(P) = 0, \\ P(T) = G, \end{cases} \tag{2.4.14}$$

satisfies the following two conditions: for almost all $s \in [t, T]$,

$$\mathcal{R}(P) \geqslant 0, \quad \mathcal{S}(P) + \mathcal{R}(P)\bar{\Theta} = 0;$$

(ii) the adapted solution (η, ζ) *to the BSDE*

$$\begin{cases} d\eta(s) = -\big[(A + B\bar{\Theta})^\top \eta + (C + D\bar{\Theta})^\top \zeta + (C + D\bar{\Theta})^\top P\sigma \\ \qquad\qquad + \bar{\Theta}^\top \rho + Pb + q\big]ds + \zeta dW(s), \quad s \in [t, T], \tag{2.4.15} \\ \eta(T) = g, \end{cases}$$

satisfies the following condition: for almost all $s \in [t, T]$,

$$B^\top \eta + D^\top \zeta + D^\top P\sigma + \rho + \mathcal{R}(P)\bar{v} = 0, \quad \text{a.s.}$$

The equivalence between Theorems 2.4.3 and 2.4.6 can be easily verified by using Proposition A.1.5 in Appendix and substituting for $\bar{\Theta}$ in (2.4.14) and (2.4.15). We leave the details to the reader.

2.5   Uniform Convexity of the Cost Functional

Recall from the functional representation (2.2.2) that the cost functional $u \mapsto J(t, x; u)$ is uniformly convex if and only if $M_2(t) \geqslant \lambda I$ for some $\lambda > 0$, which is also equivalent to

$$J^0(t, 0; u) \geqslant \lambda \mathbb{E} \int_t^T |u(s)|^2 ds, \quad \forall u \in \mathcal{U}[t, T]. \tag{2.5.1}$$

If the following *standard condition* holds for some $\delta > 0$:

$$G \geqslant 0, \quad R \geqslant \delta I, \quad Q - S^\top R^{-1} S \geqslant 0, \tag{2.5.2}$$

then the operator

$$M_2(t) = \widehat{L}_t^* G \widehat{L}_t + L_t^*(Q - S^\top R^{-1} S) L_t + (L_t^* S^\top R^{-\frac{1}{2}} + R^{\frac{1}{2}})(R^{-\frac{1}{2}} S L_t + R^{\frac{1}{2}})$$

is positive, which means that the functional $u \mapsto J^0(t, 0, u)$ is convex. The following result tells us that under (2.5.2), one actually has the uniform convexity of the cost functional.

Proposition 2.5.1 *Let (H1)–(H2) and (2.5.2) hold. Then for any $t \in [0, T)$, the mapping $u \mapsto J^0(t, 0; u)$ is uniformly convex.*

Proof Take an arbitrary $u \in \mathcal{U}[t, T]$ and let $X^{(u)}$ denote the solution of

$$\begin{cases} dX^{(u)}(s) = [A(s)X^{(u)}(s) + B(s)u(s)]ds \\ \qquad\qquad + [C(s)X^{(u)}(s) + D(s)u(s)]dW(s), \quad s \in [t, T], \\ X^{(u)}(t) = 0. \end{cases}$$

Since (2.5.2) holds, we have

$$\begin{aligned} J^0(t, 0; u) &\geqslant \mathbb{E} \int_t^T \Big[\langle QX^{(u)}, X^{(u)} \rangle + 2\langle SX^{(u)}, u \rangle + \langle Ru, u \rangle \Big] ds \\ &= \mathbb{E} \int_t^T \Big[\langle (Q - S^\top R^{-1} S) X^{(u)}, X^{(u)} \rangle + |R^{\frac{1}{2}}(u + R^{-1} SX^{(u)})|^2 \Big] ds \\ &\geqslant \delta \mathbb{E} \int_t^T |u + R^{-1} SX^{(u)}|^2 ds. \tag{2.5.3} \end{aligned}$$

Now we define a bounded linear operator $\mathfrak{L} : \mathcal{U}[t, T] \to \mathcal{U}[t, T]$ by

$$\mathfrak{L}u = u + R^{-1} SX^{(u)}.$$

It is easy to see that $\mathfrak{L}$ is bijective, and that its inverse $\mathfrak{L}^{-1}$ is given by

$$\mathfrak{L}^{-1} u = u - R^{-1} S \tilde{X}^{(u)},$$

where $\tilde{X}^{(u)}$ is the solution of

$$\begin{cases} d\tilde{X}^{(u)}(s) = \big[(A - BR^{-1}S)\tilde{X}^{(u)} + Bu\big]ds \\ \qquad\qquad + \big[(C - DR^{-1}S)\tilde{X}^{(u)} + Du\big]dW(s), \quad s \in [t, T], \\ \tilde{X}^{(u)}(t) = 0. \end{cases}$$

By the bounded inverse theorem, $\mathcal{L}^{-1}$ is bounded with $\|\mathcal{L}^{-1}\| > 0$. Thus,

$$\mathbb{E}\int_t^T |u(s)|^2 ds = \mathbb{E}\int_t^T |(\mathcal{L}^{-1}\mathcal{L}u)(s)|^2 ds \leqslant \|\mathcal{L}^{-1}\|^2\mathbb{E}\int_t^T |(\mathcal{L}u)(s)|^2 ds. \quad (2.5.4)$$

Combining (2.5.3) and (2.5.4), we obtain

$$J^0(t, 0; u) \geqslant \delta\mathbb{E}\int_t^T |(\mathcal{L}u)(s)|^2 ds \geqslant \frac{\delta}{\|\mathcal{L}^{-1}\|^2}\mathbb{E}\int_t^T |u(s)|^2 ds.$$

Since $u \in \mathcal{U}[t, T]$ is arbitrary, the desired conclusion follows. □

From Theorem 2.3.2, we see that the convexity of $u \mapsto J^0(t, 0; u)$ is necessary for the open-loop solvability of Problem (SLQ). The following result, not very surprising, says that if the convexity condition is strengthened a little, it becomes a sufficient condition.

Proposition 2.5.2 *Let (H1)–(H2) hold. Suppose that the mapping $u \mapsto J^0(0, 0; u)$ is uniformly convex. Then Problem (SLQ) is uniquely open-loop solvable, and there exists a constant $\alpha \in \mathbb{R}$ such that*

$$V^0(t, x) \geqslant \alpha|x|^2, \quad \forall(t, x) \in [0, T] \times \mathbb{R}^n. \quad (2.5.5)$$

Note that in the above, the constant α is not required to be nonnegative.

Proof of Proposition 2.5.2. By assumption, there is a constant $\lambda > 0$ such that

$$J^0(0, 0; u) \geqslant \lambda\mathbb{E}\int_0^T |u(s)|^2 ds, \quad \forall u \in \mathcal{U}[0, T]. \quad (2.5.6)$$

We claim that for any $t \in [0, T)$, (2.5.1) holds with the same constant λ. To see this, let us define the *zero-extension* of $u \in \mathcal{U}[t, T]$ as follows:

$$[01_{[0,t)} \oplus u](s) = \begin{cases} 0, & s \in [0, t), \\ u(s), & s \in [t, T]. \end{cases}$$

Clearly, $v \equiv [01_{[0,t)} \oplus u] \in \mathcal{U}[0, T]$, and due to the initial state being 0, the solution of

$$\begin{cases} dX(s) = (AX + Bv)ds + (CX + Dv)dW(s), \quad s \in [0, T], \\ X(0) = 0, \end{cases}$$

satisfies $X(s) = 0$ for all $s \in [0, t]$. Hence,

$$J^0(t, 0; u) = J^0(0, 0; [0\mathbf{1}_{[0,t)} \oplus u])$$

$$\geq \lambda \mathbb{E} \int_0^T \left|[0\mathbf{1}_{[0,t)} \oplus u](s)\right|^2 ds = \lambda \mathbb{E} \int_t^T |u(s)|^2 ds.$$

This proves our claim. Now the unique solvability of Problem (SLQ) follows by completing the square of the functional representation (2.2.2). To prove (2.5.5), we observe from Proposition 2.3.1 that for any $u \in \mathcal{U}[t, T]$,

$$J^0(t, x; u) = J^0(t, x; 0) + J^0(t, 0; u) + \mathbb{E} \int_t^T \langle \mathcal{D}_u J^0(t, x; 0)(s), u(s) \rangle ds$$

$$\geq J^0(t, x; 0) + J^0(t, 0; u) - \frac{1}{2} \mathbb{E} \int_t^T \left[\lambda |u|^2 + \frac{1}{\lambda} |\mathcal{D}_u J^0(t, x; 0)|^2 \right] ds$$

$$\geq J^0(t, x; 0) - \frac{1}{2\lambda} \mathbb{E} \int_t^T |\mathcal{D}_u J^0(t, x; 0)(s)|^2 ds.$$

Taking the infimum on the left-hand side of the above inequality over all admissible controls $u \in \mathcal{U}[t, T]$, we see that

$$V^0(t, x) \geq J^0(t, x; 0) - \frac{1}{2\lambda} \mathbb{E} \int_t^T |\mathcal{D}_u J^0(t, x; 0)(s)|^2 ds. \tag{2.5.7}$$

Then (2.5.5) follows by the fact that the functions on the right-hand side of (2.5.7) are quadratic in x and continuous in t.   □

We have seen that the convexity of the cost functional is necessary for the open-loop solvability of Problem (SLQ), while the uniform convexity is sufficient. It is natural to ask: What is the relationship between the uniform convexity of the cost functional and the closed-loop solvability of Problem (SLQ)? Recall that, roughly speaking, the closed-loop solvability of Problem (SLQ) is equivalent to the existence of a regular solution to the Riccati equation (2.4.1).

Definition 2.5.3 A solution $P \in C([t, T]; \mathbb{S}^n)$ to the Riccati equation (2.4.1) on $[t, T]$ is called *strongly regular* if there exists a constant $\lambda > 0$ such that

$$\mathcal{R}(s, P(s)) \equiv R(s) + D(s)^\top P(s) D(s) \geq \lambda I, \quad \text{a.e. } s \in [t, T]. \tag{2.5.8}$$

The Riccati equation (2.4.1) is called *strongly regularly solvable* if it admits a strongly regular solution.

Remark 2.5.4 Clearly, (2.5.8) implies the conditions (i)–(iii) of Definition 2.4.1. Thus, a strongly regular solution is also regular. Moreover, by Corollary 2.4.5, if a strongly regular solution exists, it must be unique.

Now we summarize the relevant results concerning Problem (SLQ) in the following diagram:

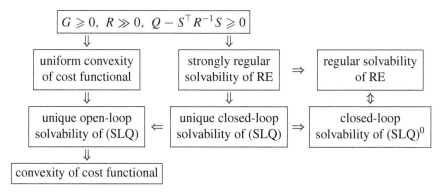

where "RE" stands for the Riccati equation (2.4.1). It is obvious that the uniform convexity of the cost functional does not imply the standard condition (2.5.2). Therefore, it is a desire to establish the following:

To achieve this, we first present the following proposition, which will play a key technical role later.

Proposition 2.5.5 *Let (H1)–(H2) hold. Let $\Theta \in \mathbf{\Theta}[0, T]$, and let $P \in C([0, T]; \mathbb{S}^n)$ be the solution to the Lyapunov equation*

$$\begin{cases} \dot{P} + P(A + B\Theta) + (A + B\Theta)^\top P + (C + D\Theta)^\top P(C + D\Theta) \\ \quad + \Theta^\top R\Theta + S^\top \Theta + \Theta^\top S + Q = 0, \quad a.e.\ s \in [0, T], \\ P(T) = G. \end{cases}$$

If (2.5.6) holds for some constant $\lambda > 0$, then

$$\mathcal{R}(s, P(s)) \geqslant \lambda I, \qquad a.e.\ s \in [0, T], \tag{2.5.9}$$
$$P(s) \geqslant \alpha I, \qquad \forall s \in [0, T], \tag{2.5.10}$$

where α is the constant in (2.5.5).

Proof For fixed but arbitrary $(t, x) \in [0, T) \times \mathbb{R}^n$ and $u \in \mathcal{U}[t, T]$, let X denote the solution to the closed-loop system

$$\begin{cases} dX(s) = [(A + B\Theta)X + Bu]ds + [(C + D\Theta)X + Du]dW, \quad s \in [t, T], \\ X(t) = x. \end{cases}$$

Applying Itô's formula to $s \mapsto \langle P(s)X(s), X(s) \rangle$ yields

$$
\begin{aligned}
&\mathbb{E}\langle GX(T), X(T) \rangle - \langle P(t)x, x \rangle \\
&= \mathbb{E} \int_t^T \Big[-\langle (\Theta^\top R\Theta + S^\top \Theta + \Theta^\top S + Q)X, X \rangle \\
&\quad + 2\langle (B^\top P + D^\top PC + D^\top PD\Theta)X, u \rangle + \langle D^\top PDu, u \rangle \Big] ds.
\end{aligned}
$$

Substituting for $\mathbb{E}\langle GX(T), X(T) \rangle$ in $J^0(t, x; \Theta X + u)$ gives

$$
\begin{aligned}
J^0(t, x; \Theta X + u) &= \langle P(t)x, x \rangle + \mathbb{E} \int_t^T \Big\{ 2\langle [\mathcal{S}(P) + \mathcal{R}(P)\Theta]X, u \rangle \\
&\quad + \langle \mathcal{R}(P)u, u \rangle \Big\} ds.
\end{aligned}
\tag{2.5.11}
$$

To prove (2.5.9), we take the initial pair (t, x) to be $(0, 0)$. If (2.5.6) holds, then

$$
\begin{aligned}
\lambda \mathbb{E} \int_0^T |\Theta X + u|^2 ds &\leqslant J^0(0, 0; \Theta X + u) \\
&= \mathbb{E} \int_0^T \Big\{ 2\langle [\mathcal{S}(P) + \mathcal{R}(P)\Theta]X, u \rangle + \langle \mathcal{R}(P)u, u \rangle \Big\} ds,
\end{aligned}
$$

which implies that

$$
\begin{aligned}
&\mathbb{E} \int_0^T \Big(2\langle \{\mathcal{S}(P) + [\mathcal{R}(P) - \lambda I]\Theta\}X, u \rangle + \langle [\mathcal{R}(P) - \lambda I]u, u \rangle \Big) ds \\
&= \lambda \mathbb{E} \int_0^T |\Theta(s)X(s)|^2 ds \geqslant 0.
\end{aligned}
\tag{2.5.12}
$$

Fix an arbitrary $u_0 \in \mathbb{R}^m$ and take $u(s) = u_0 \mathbf{1}_{[r, r+h]}(s)$, with $0 \leqslant r < r + h \leqslant T$. Then

$$
\begin{cases}
\dfrac{d}{ds}[\mathbb{E}X(s)] = [A(s) + B(s)\Theta(s)]\mathbb{E}X(s) + B(s)u_0 \mathbf{1}_{[r, r+h]}(s), & s \in [0, T], \\
\mathbb{E}X(0) = 0.
\end{cases}
$$

By the variation of constants formula, we have

$$
\mathbb{E}X(s) = \begin{cases}
0, & s \in [0, r], \\
\Phi(s) \displaystyle\int_r^{s \wedge (r+h)} \Phi(v)^{-1} B(v)u_0 dv, & s \in [r, T],
\end{cases}
$$

where Φ is the solution to the following matrix ODE:

$$\begin{cases} \dot{\Phi}(s) = [A(s) + B(s)\Theta(s)]\Phi(s), & s \in [0, T], \\ \Phi(0) = I. \end{cases}$$

Consequently, with

$$M(s) \triangleq \left\{ \mathcal{S}(s, P(s)) + [\mathcal{R}(s, P(s)) - \lambda I]\Theta(s) \right\} \Phi(s) \int_r^s \Phi(v)^{-1} B(v) dv,$$

Equation (2.5.12) becomes

$$\int_r^{r+h} \left\{ 2\langle M(s)u_0, u_0 \rangle + \langle [\mathcal{R}(s, P(s)) - \lambda I]u_0, u_0 \rangle \right\} ds \geq 0.$$

Dividing both sides of the above by h and letting $h \to 0$, we obtain

$$\langle [\mathcal{R}(r, P(r)) - \lambda I]u_0, u_0 \rangle \geq 0, \quad \text{a.e. } r \in [0, T].$$

Since $u_0 \in \mathbb{R}^m$ is arbitrary, (2.5.9) follows. To prove (2.5.10), we still let $(t, x) \in [0, T) \times \mathbb{R}^n$ be arbitrary but take $u = 0$, Then by Proposition 2.5.2 and (2.5.11), we have

$$\alpha |x|^2 \leq V^0(t, x) \leq J^0(t, x; \Theta X + 0) = \langle P(t)x, x \rangle,$$

and (2.5.10) follows since (t, x) is arbitrary. $\square$

We now present the main result of this section.

Theorem 2.5.6 *Let (H1)–(H2) hold. Then the following are equivalent:*

(i) the mapping $u \mapsto J^0(0, 0; u)$ is uniformly convex;
(ii) the Riccati equation (2.4.1) admits a strongly regular solution on $[0, T]$;
(iii) there exists an $\mathbb{S}^n$-valued function P such that (2.5.9) holds for some constant $\lambda > 0$ and $V^0(t, x) = \langle P(t)x, x \rangle$ for all $(t, x) \in [0, T] \times \mathbb{R}^n$.

Proof (i) $\Rightarrow$ (ii): We may assume that (2.5.6) holds for some constant $\lambda > 0$. Let P_0 be the solution to the Lyapunov equation

$$\begin{cases} \dot{P}_0 + P_0 A + A^\top P_0 + C^\top P_0 C + Q = 0, & \text{a.e. } s \in [0, T], \\ P_0(T) = G. \end{cases}$$

Applying Proposition 2.5.5 with $\Theta = 0$, we obtain that

$$R(s) + D(s)^\top P_0(s)D(s) \geq \lambda I, \quad P_0(s) \geq \alpha I, \quad \text{a.e. } s \in [0, T].$$

Next, inductively, for $i = 0, 1, 2, \cdots$, we set

$$
\begin{cases}
\Theta_i = -(R + D^\top P_i D)^{-1}(B^\top P_i + D^\top P_i C + S), \\
A_i = A + B\Theta_i, \quad C_i = C + D\Theta_i,
\end{cases}
\tag{2.5.13}
$$

and let P_{i+1} be the solution to the following Lyapunov equation on $[0, T]$:

$$
\begin{cases}
\dot{P}_{i+1} + P_{i+1} A_i + A_i^\top P_{i+1} + C_i^\top P_{i+1} C_i + \Theta_i^\top R \Theta_i + S^\top \Theta_i + \Theta_i^\top S + Q = 0, \\
P_{i+1}(T) = G.
\end{cases}
$$

Again by Proposition 2.5.5, we have for almost all $s \in [0, T]$,

$$
R(s) + D(s)^\top P_{i+1}(s) D(s) \geqslant \lambda I, \quad P_{i+1}(s) \geqslant \alpha I.
\tag{2.5.14}
$$

We claim that $\{P_i\}_{i=1}^\infty$ converges pointwise to a limit P that is a strongly regular solution to the Riccati equation (2.4.1). To prove this, we set

$$
\Delta_i = P_i - P_{i+1}, \quad \Lambda_i = \Theta_{i-1} - \Theta_i; \quad i \geqslant 1.
$$

For $i \geqslant 1$, we have

$$
\begin{aligned}
-\dot{\Delta}_i &= \dot{P}_{i+1} - \dot{P}_i \\
&= P_i A_{i-1} + A_{i-1}^\top P_i + C_{i-1}^\top P_i C_{i-1} + \Theta_{i-1}^\top R \Theta_{i-1} \\
&\quad + S^\top \Theta_{i-1} + \Theta_{i-1}^\top S - P_{i+1} A_i - A_i^\top P_{i+1} \\
&\quad - C_i^\top P_{i+1} C_i - \Theta_i^\top R \Theta_i - S^\top \Theta_i - \Theta_i^\top S \\
&= \Delta_i A_i + A_i^\top \Delta_i + C_i^\top \Delta_i C_i + P_i (A_{i-1} - A_i) \\
&\quad + (A_{i-1} - A_i)^\top P_i + C_{i-1}^\top P_i C_{i-1} - C_i^\top P_i C_i \\
&\quad + \Theta_{i-1}^\top R \Theta_{i-1} - \Theta_i^\top R \Theta_i + S^\top \Lambda_i + \Lambda_i^\top S.
\end{aligned}
\tag{2.5.15}
$$

By (2.5.13), it is easy to see that

$$
\begin{cases}
A_{i-1} - A_i = B\Lambda_i, \\
C_{i-1}^\top P_i C_{i-1} - C_i^\top P_i C_i = \Lambda_i^\top D^\top P_i D\Lambda_i + C_i^\top P_i D\Lambda_i + \Lambda_i^\top D^\top P_i C_i, \\
\Theta_{i-1}^\top R \Theta_{i-1} - \Theta_i^\top R \Theta_i = \Lambda_i^\top R\Lambda_i + \Lambda_i^\top R\Theta_i + \Theta_i^\top R\Lambda_i.
\end{cases}
\tag{2.5.16}
$$

Note also that

$$
B^\top P_i + D^\top P_i C_i + R\Theta_i + S = B^\top P_i + D^\top P_i C + S + (R + D^\top P_i D)\Theta_i = 0.
$$

Then plugging (2.5.16) into (2.5.15) yields

$$
\begin{aligned}
&- (\dot{\Delta}_i + \Delta_i A_i + A_i^\top \Delta_i + C_i^\top \Delta_i C_i) \\
&= P_i B \Lambda_i + \Lambda_i^\top B^\top P_i + \Lambda_i^\top D^\top P_i D \Lambda_i + C_i^\top P_i D \Lambda_i + \Lambda_i^\top D^\top P_i C_i \\
&\quad + \Lambda_i^\top R \Lambda_i + \Lambda_i^\top R \Theta_i + \Theta_i^\top R \Lambda_i + S^\top \Lambda_i + \Lambda_i^\top S \\
&= \Lambda_i^\top (R + D^\top P_i D) \Lambda_i + (P_i B + C_i^\top P_i D + \Theta_i^\top R + S^\top) \Lambda_i \\
&\quad + \Lambda_i^\top (B^\top P_i + D^\top P_i C_i + R \Theta_i + S) \\
&= \Lambda_i^\top (R + D^\top P_i D) \Lambda_i.
\end{aligned} \tag{2.5.17}
$$

Let Φ_i be the solution to (2.2.1) with (A, C) replaced by (A_i, C_i), and denote

$$
Q_i \triangleq \Lambda_i^\top (R + D^\top P_i D) \Lambda_i.
$$

Repeating the argument that led to (2.2.7) and noting $\Delta_i(T) = 0$, we obtain

$$
\Delta_i(t) = \mathbb{E} \int_t^T \left[\Phi_i(s) \Phi_i(t)^{-1} \right]^\top Q_i(s) \left[\Phi_i(s) \Phi_i(t)^{-1} \right] ds, \quad t \in [0, T].
$$

By (2.5.14), $Q_i(s) \geqslant 0$, we see that $\Delta_i(t) \geqslant 0$ for all $t \in [0, T]$ and hence

$$
P_1(s) \geqslant P_i(s) \geqslant P_{i+1}(s) \geqslant \alpha I, \quad \forall s \in [0, T], \ \forall i \geqslant 1. \tag{2.5.18}
$$

It follows from the monotone convergence theorem that the limit $P(s) \equiv \lim_{i \to \infty} P_i(s)$ exists for all $s \in [0, T]$. To show that the limit function P is a strongly regular solution to the Riccati equation (2.4.1) over $[0, T]$, we observe first that

$$
\begin{aligned}
P_{i+1}(t) = G + \int_t^T \Big(&P_{i+1} A_i + A_i^\top P_{i+1} + C_i^\top P_{i+1} C_i + \Theta_i^\top R \Theta_i \\
&+ S^\top \Theta_i + \Theta_i^\top S + Q \Big) ds.
\end{aligned} \tag{2.5.19}
$$

By (2.5.14), we have for almost all $s \in [0, T]$,

$$
R(s) + D(s)^\top P(s) D(s) = \lim_{i \to \infty} \left[R(s) + D(s)^\top P_i(s) D(s) \right] \geqslant \lambda I,
$$

and as $i \to \infty$, we have a.e.

$$
\begin{aligned}
\Theta_i &\to -(R + D^\top P D)^{-1} (B^\top P + D^\top P C + S) \equiv \Theta, \\
A_i &\to A + B\Theta, \qquad C_i \to C + D\Theta.
\end{aligned}
$$

Also, according to (2.5.18), the sequence $\{P_i\}_{i=1}^{\infty}$ is uniformly bounded. We now let $i \to \infty$ in (2.5.19), using the dominated convergence theorem, to obtain

$$P(t) = G + \int_t^T \Big[P(A + B\Theta) + (A + B\Theta)^\top P + (C + D\Theta)^\top P(C + D\Theta)$$
$$+ \Theta^\top R\Theta + S^\top \Theta + \Theta^\top S + Q \Big] ds.$$

By differentiating both sides of the above and substituting for Θ, we see that P satisfies the Riccati equation (2.4.1).

(ii) $\Rightarrow$ (i): Let P be the strongly regular solution to the Riccati equation (2.4.1) on $[0, T]$, and set

$$\Theta \triangleq -\mathcal{R}(P)^{-1}\mathcal{S}(P) \in L^2(0, T; \mathbb{R}^{m \times n}).$$

For any $u \in \mathcal{U}[0, T]$, let $X^{(u)}$ denote the solution of

$$\begin{cases} dX^{(u)}(s) = \big(AX^{(u)} + Bu\big)ds + \big(CX^{(u)} + Du\big)dW(s), & s \in [0, T], \\ X^{(u)}(0) = 0. \end{cases}$$

Applying Itô's formula to $s \mapsto \langle P(s)X^{(u)}(s), X^{(u)}(s) \rangle$ yields

$$\mathbb{E}\langle GX^{(u)}(T), X^{(u)}(T) \rangle = \mathbb{E}\int_0^T \Big[\langle (\dot{P} + PA + A^\top P + C^\top PC)X^{(u)}, X^{(u)} \rangle$$
$$+ 2\langle (B^\top P + D^\top PC)X^{(u)}, u \rangle + \langle D^\top PDu, u \rangle \Big] ds.$$

Substituting for $\mathbb{E}\langle GX^{(u)}(T), X^{(u)}(T) \rangle$ in $J^0(0, 0; u)$ gives

$$J^0(0, 0; u) = \mathbb{E}\int_0^T \Big[\langle (\dot{P} + \mathcal{Q}(P))X^{(u)}, X^{(u)} \rangle + 2\langle \mathcal{S}(P)X^{(u)}, u \rangle + \langle \mathcal{R}(P)u, u \rangle \Big] ds$$
$$= \mathbb{E}\int_0^T \langle \mathcal{R}(P)(u - \Theta X^{(u)}), u - \Theta X^{(u)} \rangle ds$$
$$\geqslant \lambda \mathbb{E}\int_0^T |u - \Theta X^{(u)}|^2 ds,$$

for some constant $\lambda > 0$. Now we can repeat the argument employed in the proof of Proposition 2.5.1, replacing $R^{-1}S$ by $-\Theta$, to conclude that

$$\mathbb{E}\int_0^T |u(s) - \Theta(s)X^{(u)}(s)|^2 ds \geqslant \delta \mathbb{E}\int_0^T |u(s)|^2 ds$$

for some constant $\delta > 0$. The uniform convexity of $u \mapsto J^0(0, 0; u)$ follows readily.

(ii) $\Rightarrow$ (iii): This is an immediate consequence of Theorem 2.4.3.

(iii) $\Rightarrow$ (ii): Let us consider, for any $\varepsilon > 0$, the cost functional

$$J_\varepsilon^0(t, x; u) = J^0(t, x; u) + \varepsilon \mathbb{E} \int_t^T |u(s)|^2 ds.$$

Since Problem (SLQ)^0 is finite by assumption, it follows from Proposition 2.2.6 that the mapping $u \mapsto J_\varepsilon^0(0, 0; u)$ is uniformly convex. Then the implication (i) $\Rightarrow$ (ii) suggests that the Riccati equation

$$\begin{cases} \dot{P}_\varepsilon + \mathcal{Q}(P_\varepsilon) - \mathcal{S}(P_\varepsilon)^\top [\mathcal{R}(P_\varepsilon) + \varepsilon I]^{-1} \mathcal{S}(P_\varepsilon) = 0, & s \in [0, T], \\ P_\varepsilon(T) = G, \end{cases}$$

admits a (unique) strongly regular solution P_ε. According to Theorem 2.4.3 (noting that for Problem (SLQ)^0, the adapted solution (η, ζ) to the BSDE (2.4.3) is identically zero), we have

$$V_\varepsilon^0(t, x) \triangleq \inf_{u \in \mathcal{U}[t,T]} J_\varepsilon^0(t, x; u) = \langle P_\varepsilon(t)x, x \rangle.$$

By the proof of Proposition 2.2.6, we know that

$$P_\varepsilon(t) \searrow P(t) \quad \text{as } \varepsilon \searrow 0, \quad \forall t \in [0, T],$$

and by assumption (2.5.9), we have

$$\mathcal{R}(s, P_\varepsilon(s)) \geqslant \mathcal{R}(s, P(s)) \geqslant \lambda I, \quad \text{a.e. } s \in [0, T], \forall \varepsilon > 0.$$

Thus, we may apply the dominated convergence theorem to obtain that

$$\begin{aligned} P(t) = \lim_{\varepsilon \to 0} P_\varepsilon(t) &= G + \lim_{\varepsilon \to 0} \int_t^T \left\{ \mathcal{Q}(P_\varepsilon) - \mathcal{S}(P_\varepsilon)^\top [\mathcal{R}(P_\varepsilon) + \varepsilon I]^{-1} \mathcal{S}(P_\varepsilon) \right\} ds \\ &= G + \int_t^T \left\{ \mathcal{Q}(P) - \mathcal{S}(P)^\top \mathcal{R}(P)^{-1} \mathcal{S}(P) \right\} ds, \end{aligned}$$

which implies that P is a strongly regular solution of (2.4.1). $\qquad\square$

Combining Theorem 2.4.3, Proposition 2.5.2, and Theorem 2.5.6, we obtain the following corollary.

Corollary 2.5.7 *Let (H1)–(H2) hold. Suppose that the mapping $u \mapsto J^0(0, 0; u)$ is uniformly convex. Then Problem (SLQ) is both uniquely closed-loop solvable and uniquely open-loop solvable. The unique closed-loop optimal strategy $(\bar{\Theta}, \bar{v})$ over $[t, T]$ is given by*

$$\bar{\Theta} = -\mathcal{R}(P)^{-1} \mathcal{S}(P), \quad \bar{v} = -\mathcal{R}(P)^{-1}(B^\top \eta + D^\top \zeta + D^\top P\sigma + \rho),$$

where P is the unique strongly regular solution to the Riccati equation (2.4.1) *over* [0, T] *and* (η, ζ) *is the adapted solution to the BSDE* (2.4.3). *The unique open-loop optimal control* $\bar{u}$ *for the initial pair* (t, x) *is given by*

$$\bar{u} = \bar{\Theta}\bar{X} + \bar{v},$$

where $\bar{X}$ *is the solution to the closed-loop system*

$$\begin{cases} d\bar{X}(s) = \big[(A + B\bar{\Theta})\bar{X} + B\bar{v} + b\big]ds \\ \qquad\quad + \big[(C + D\bar{\Theta})\bar{X} + D\bar{v} + \sigma\big]dW(s), \quad s \in [t, T], \\ \bar{X}(t) = x. \end{cases}$$

Proof According to Theorem 2.5.6, the Riccati equation (2.4.1) admits a unique strongly regular solution $P \in C([0, T]; \mathbb{S}^n)$. Since the strongly regular solution P satisfies

$$R(s) + D(s)^\top P(s)D(s) \geqslant \lambda I, \quad \text{a.e. } s \in [0, T]$$

for some $\lambda > 0$, the adapted solution (η, ζ) to the BSDE (2.4.3) automatically satisfies the conditions (2.4.4) and (2.4.5). Now applying Theorem 2.4.3 and noting that $R + D^\top PD$ is invertible, we obtain the unique closed-loop solvability of Problem (SLQ). The rest follows directly from Proposition 2.5.2 and the fact that the outcome of a closed-loop optimal strategy is open-loop optimal. □

Remark 2.5.8 If we consider Problem (SLQ)^0 instead of Problem (SLQ), then the process $\bar{v}$ in Corollary 2.5.7 is identically zero.

Theorem 2.5.6 establishes the equivalence between the uniform convexity of $u \mapsto J^0(0, 0; u)$ and the strongly regular solvability of (2.4.1). Clearly, some easy checkable sufficient conditions for the uniform convexity of $u \mapsto J^0(t, 0; u)$ are desirable. The following is such a result.

Theorem 2.5.9 *Let (H1)–(H2) hold. Let* Π *be the solution to the Lyapunov equation*

$$\begin{cases} \dot{\Pi} + \Pi A + A^\top \Pi + C^\top \Pi C + Q - Q_0 = 0, \quad s \in [t, T], \\ \Pi(T) = G, \end{cases} \quad (2.5.20)$$

for some $Q_0 \in L^1(t, T; \mathbb{S}_+^n)$. *If for some* $\delta > 0$,

$$\Delta(\Pi) \triangleq \mathcal{R}(\Pi) - \mathcal{S}(\Pi)Q_0^{-1}\mathcal{S}(\Pi)^\top \geqslant \delta I_m, \quad \text{a.e. on } [t, T], \quad (2.5.21)$$

then $u \mapsto J^0(t, 0; u)$ *is uniformly convex.*

Proof Fix an arbitrary $u \in \mathcal{U}[t, T]$ and let X be the solution to

$$
\begin{cases}
dX(s) = (AX + Bu)ds + (CX + Du)dW(s), & s \in [t, T], \\
X(t) = 0.
\end{cases}
$$

Then with the notation

$$
\Gamma \triangleq -(\Pi A + A^\top \Pi + C^\top \Pi C + Q - Q_0),
$$

we have by Itô's formula that

$$
d(\Pi X) = (\Gamma X + \Pi A X + \Pi B u)ds + (\Pi C X + \Pi D u)dW,
$$

and hence

$$
\begin{aligned}
\mathbb{E}\langle GX(T), X(T)\rangle &= \mathbb{E}\langle \Pi(T)X(T), X(T)\rangle - \mathbb{E}\langle \Pi(t)X(t), X(t)\rangle \\
&= \mathbb{E}\int_t^T \Big[\langle \Gamma X + \Pi A X + \Pi B u, X\rangle + \langle \Pi X, AX + Bu\rangle \\
&\quad + \langle \Pi C X + \Pi D u, CX + Du\rangle\Big]ds \\
&= \mathbb{E}\int_t^T \Big[\langle(Q_0 - Q)X, X\rangle + 2\langle(B^\top \Pi + D^\top \Pi C)X, u\rangle \\
&\quad + \langle D^\top \Pi D u, u\rangle\Big]ds.
\end{aligned}
$$

Substituting the above into the cost functional, we obtain

$$
\begin{aligned}
J^0(t, 0; u) &= \mathbb{E}\int_t^T \Big[\langle Q_0 X, X\rangle + 2\langle \mathcal{S}(\Pi)X, u\rangle + \langle \mathcal{R}(\Pi)u, u\rangle\Big]ds \\
&= \mathbb{E}\int_t^T \Big[|Q_0^{\frac{1}{2}} X + Q_0^{-\frac{1}{2}}\mathcal{S}(\Pi)^\top u|^2 + \langle \Delta(\Pi)u, u\rangle\Big]ds \\
&\geqslant \delta \mathbb{E}\int_t^T |u(s)|^2 ds.
\end{aligned}
$$

This proves our conclusion. $\square$

The above result gives some compatibility conditions among the coefficients of the state equation and the weighting matrices in the cost functional that ensure the uniform convexity of $u \mapsto J^0(t, 0; u)$. We now look at some special cases.

(i) Let $\lambda > 0$ and $Q_0 = \lambda I_n$. Then with Π_λ denoting the solution of

$$
\begin{cases}
\dot{\Pi}_\lambda + \Pi_\lambda A + A^\top \Pi_\lambda + C^\top \Pi_\lambda C + Q - \lambda I_n = 0, & s \in [t, T], \\
\Pi_\lambda(T) = G,
\end{cases}
$$

the corresponding condition (2.5.21) reads

$$\mathcal{R}(\Pi_\lambda) - \lambda^{-1}\mathcal{S}(\Pi_\lambda)\mathcal{S}(\Pi_\lambda)^\top \geqslant \delta I_m, \quad \text{a.e. on } [t, T].$$

(ii) When $B = 0$, $C = 0$, $S = 0$, (2.5.20) becomes

$$\begin{cases} \dot{\Pi} + \Pi A + A^\top \Pi + Q - Q_0 = 0, \quad s \in [t, T], \\ \Pi(T) = G, \end{cases}$$

and (2.5.21) simply reads $\mathcal{R}(\Pi) \geqslant \delta I_m$ a.e. on $[t, T]$.

2.6 Finiteness and Solvability Under Other Conditions

From Corollary 2.2.4 and Example 2.2.5, we know that the convexity of $u \mapsto J(t, x; u)$, or equivalently, $M_2(t) \geqslant 0$, is necessary for Problem (SLQ) to be finite at (t, x), but not sufficient. Recall Problem $(SLQ)_\varepsilon$ $(\varepsilon > 0)$ introduced at the end of Sect. 2.2, for which the cost functional is defined by

$$J_\varepsilon(t, x; u) \triangleq J(t, x; u) + \varepsilon\mathbb{E} \int_t^T |u(s)|^2 ds. \tag{2.6.1}$$

When the following necessary condition for the finiteness of Problem (SLQ) at t holds:

$$M_2(t) \geqslant 0, \tag{2.6.2}$$

the mapping $u \mapsto J_\varepsilon(t, x; u)$ is uniformly convex, and the results from the previous section apply to Problem $(SLQ)_\varepsilon$. For simplicity, we take $t = 0$ in (2.6.2), that is, we assume

$$M_2(0) \geqslant 0. \tag{2.6.3}$$

Then the Riccati equation

$$\begin{cases} \dot{P}_\varepsilon + \mathcal{Q}(P_\varepsilon) - \mathcal{S}(P_\varepsilon)^\top [\mathcal{R}(P_\varepsilon) + \varepsilon I]^{-1}\mathcal{S}(P_\varepsilon) = 0, \quad s \in [0, T], \\ P_\varepsilon(T) = G \end{cases} \tag{2.6.4}$$

admits a unique strongly regular solution P_ε, and the BSDE

$$\begin{cases} d\eta_\varepsilon(s) = -\big[(A + B\Theta_\varepsilon)^\top \eta_\varepsilon + (C + D\Theta_\varepsilon)^\top \zeta_\varepsilon + (C + D\Theta_\varepsilon)^\top P_\varepsilon \sigma \\ \qquad\qquad + \Theta_\varepsilon^\top \rho + P_\varepsilon b + q\big]ds + \zeta_\varepsilon dW, \quad s \in [0, T], \\ \eta_\varepsilon(T) = g \end{cases} \tag{2.6.5}$$

admits a unique adapted solution $(\eta_\varepsilon, \zeta_\varepsilon)$, where

$$\Theta_\varepsilon(s) \triangleq -[\mathcal{R}(s, P_\varepsilon(s)) + \varepsilon I]^{-1} S(s, P_\varepsilon(s)), \quad s \in [0, T]. \tag{2.6.6}$$

Let $v_\varepsilon : [0, T] \times \Omega \to \mathbb{R}^m$ be defined by

$$v_\varepsilon \triangleq -[\mathcal{R}(P_\varepsilon) + \varepsilon I]^{-1} (B^\top \eta_\varepsilon + D^\top \zeta_\varepsilon + D^\top P_\varepsilon \sigma + \rho). \tag{2.6.7}$$

For an initial pair $(t, x) \in [0, T] \times \mathbb{R}^n$, denote by $X_\varepsilon = \{X_\varepsilon(s); t \leqslant s \leqslant T\}$ the solution to the closed-loop system

$$\begin{cases} dX_\varepsilon(s) = [(A + B\Theta_\varepsilon)X_\varepsilon + Bv_\varepsilon + b]ds \\ \qquad\qquad + [(C + D\Theta_\varepsilon)X_\varepsilon + Dv_\varepsilon + \sigma]dW, \quad s \in [t, T], \\ X_\varepsilon(t) = x. \end{cases} \tag{2.6.8}$$

Then according to Corollary 2.5.7, the control process u_ε defined by

$$u_\varepsilon(s) \triangleq \Theta_\varepsilon(s)X_\varepsilon(s) + v_\varepsilon(s), \quad s \in [t, T], \tag{2.6.9}$$

is the unique open-loop optimal control of Problem $(SLQ)_\varepsilon$ for (t, x). In particular, for Problem $(SLQ)_\varepsilon^0$, we have

$$V_\varepsilon^0(t, x) = \langle P_\varepsilon(t)x, x \rangle, \quad \forall (t, x) \in [0, T] \times \mathbb{R}^n. \tag{2.6.10}$$

Moreover, from (2.6.1) and the representation (2.2.4), it is not hard to see that

$$P_{\varepsilon_1}(t) \leqslant P_{\varepsilon_2}(t) \leqslant M_0(t), \quad \forall t \in [0, T], \forall 0 < \varepsilon_1 \leqslant \varepsilon_2.$$

Theorem 2.6.1 *Let (H1)–(H2) hold and assume (2.6.3).*

(i) *If Problem (SLQ) is finite at (t, x), then $\{u_\varepsilon\}_{\varepsilon>0}$ defined by (2.6.9) is a minimizing family of $u \mapsto J(t, x; u)$, i.e.,*

$$\lim_{\varepsilon \to 0} J(t, x; u_\varepsilon) = \inf_{u \in \mathcal{U}[t,T]} J(t, x; u) = V(t, x).$$

(ii) *The following statements are equivalent:*

    (a) *Problem $(SLQ)^0$ is finite at time $t = 0$;*
    (b) *Problem $(SLQ)^0$ is finite at all $s \in [0, T]$;*
    (c) *$\{P_\varepsilon(0)\}_{\varepsilon>0}$ is bounded from below.*

In this case, $P(s) \equiv \lim_{\varepsilon \to 0} P_\varepsilon(s)$ exists for all $s \in [0, T]$. Moreover,

$$V^0(s, x) = \langle P(s)x, x \rangle, \qquad \forall (s, x) \in [0, T] \times \mathbb{R}^n, \qquad (2.6.11)$$

$$\mathcal{R}(s, P(s)) \geq 0, \qquad \text{a.e. } s \in [0, T], \qquad (2.6.12)$$

$$N(s) \leq P(s) \leq M_0(s), \qquad \forall s \in [0, T], \qquad (2.6.13)$$

where M_0 is the solution to the Lyapunov equation (2.2.6) and N is a continuous function given by

$$N(s) = \left[\Phi_A(s)^{-1} \right]^\top \left\{ P(0) - \int_0^s \left[\Phi_A^\top (C^\top M_0 C + Q) \Phi_A \right](r) dr \right\} \Phi_A(s)^{-1},$$

with Φ_A being the solution to the matrix ODE

$$\begin{cases} \dot{\Phi}_A(s) = A(s)\Phi_A(s), & s \in [0, T], \\ \Phi_A(0) = I_n. \end{cases}$$

Proof (i) This follows directly from Proposition 1.3.2 of Chap. 1.

(ii) The implication $(b) \Rightarrow (a)$ is obvious.

For the implication $(a) \Rightarrow (c)$, we note first that by Proposition 2.2.6(ii), there exists a matrix $P(0) \in \mathbb{S}^n$ such that

$$V^0(0, x) = \langle P(0)x, x \rangle, \qquad \forall x \in \mathbb{R}^n. \qquad (2.6.14)$$

By Proposition 2.2.6(i) and (2.6.10),

$$\lim_{\varepsilon \to 0} \langle P_\varepsilon(0)x, x \rangle = \lim_{\varepsilon \to 0} V_\varepsilon^0(0, x) = V^0(0, x) = \langle P(0)x, x \rangle, \qquad \forall x \in \mathbb{R}^n.$$

Since x is arbitrary, we have $\lim_{\varepsilon \to 0} P_\varepsilon(0) = P(0)$. In particular, as a decreasing, convergent family, $\{P_\varepsilon(0)\}_{\varepsilon > 0}$ must be bounded from below.

It remains to show the implication $(c) \Rightarrow (b)$. To this end, let $\beta \in \mathbb{R}$ be such that $P_\varepsilon(0) \geq \beta I$ for all $\varepsilon > 0$. Then for any $x \in \mathbb{R}^n$ and $u \in \mathcal{U}[0, T]$,

$$J^0(0, x; u) + \varepsilon \mathbb{E} \int_0^T |u(s)|^2 ds \geq V_\varepsilon^0(0, x) = \langle P_\varepsilon(0)x, x \rangle \geq \beta |x|^2, \qquad \forall \varepsilon > 0.$$

Letting $\varepsilon \to 0$ in the above, we obtain

$$J^0(0, x; u) \geq \beta |x|^2, \qquad \forall x \in \mathbb{R}^n, \ \forall u \in \mathcal{U}[0, T],$$

which implies the finiteness of Problem (SLQ)^0 at time $t = 0$. By Proposition 2.2.6(ii), we can find a $P(0) \in \mathbb{S}^n$ such that (2.6.14) holds. Since for every $\varepsilon > 0$,

$$\langle P(0)x, x \rangle = V^0(0, x) \leq V_\varepsilon^0(0, x) = \langle P_\varepsilon(0)x, x \rangle, \qquad \forall x \in \mathbb{R}^n,$$

we have

$$P(0) \leqslant P_\varepsilon(0), \quad \forall \varepsilon > 0. \tag{2.6.15}$$

By the representation (2.2.4), for any $\varepsilon > 0$, $s \in [0, T]$, and $x \in \mathbb{R}^n$,

$$\langle P_\varepsilon(s)x, x \rangle = V_\varepsilon^0(s, x) \leqslant J_\varepsilon^0(s, x; 0) = J^0(s, x; 0) = \langle M_0(s)x, x \rangle,$$

which leads to

$$P_\varepsilon(s) \leqslant M_0(s), \quad \forall s \in [0, T], \ \forall \varepsilon > 0. \tag{2.6.16}$$

On the other hand, setting

$$\Pi_\varepsilon \triangleq \mathcal{S}(P_\varepsilon)^\top [\mathcal{R}(P_\varepsilon) + \varepsilon I]^{-1} \mathcal{S}(P_\varepsilon)$$

and differentiating $s \mapsto \Phi_A(s)^\top P_\varepsilon(s) \Phi_A(s)$ yield

$$\frac{d}{ds} \Big[\Phi_A(s)^\top P_\varepsilon(s) \Phi_A(s) \Big] = \Phi_A(s)^\top \Big[\Pi_\varepsilon(s) - C(s)^\top P_\varepsilon(s) C(s) - Q(s) \Big] \Phi_A(s).$$

Thus, by combining (2.6.15)–(2.6.16) and noting that $\Pi_\varepsilon \geqslant 0$, we obtain

$$\Phi_A(s)^\top P_\varepsilon(s) \Phi_A(s) = P_\varepsilon(0) + \int_0^s \Phi_A^\top \big(\Pi_\varepsilon - C^\top P_\varepsilon C - Q \big) \Phi_A dr$$

$$\geqslant P(0) - \int_0^s \Phi_A^\top \big(C^\top M_0 C + Q \big) \Phi_A dr.$$

Pre- and post-multiplying the above by $[\Phi_A(s)^{-1}]^\top$ and $\Phi_A(s)^{-1}$ respectively yields

$$P_\varepsilon(s) \geqslant N(s), \quad \forall s \in [0, T], \ \forall \varepsilon > 0. \tag{2.6.17}$$

Therefore, for any $x \in \mathbb{R}^n$,

$$V^0(s, x) = \lim_{\varepsilon \to 0} V_\varepsilon^0(s, x) = \lim_{\varepsilon \to 0} \langle P_\varepsilon(s)x, x \rangle \geqslant \langle N(s)x, x \rangle > -\infty.$$

This gives the finiteness of Problem (SLQ)^0 at $s \in [0, T]$.

Finally, the monotone convergence theorem implies that the limit $P(s) \equiv \lim_{\varepsilon \to 0} P_\varepsilon(s)$ exists for all $s \in [0, T]$, and the inequalities (2.6.13) follow immediately from (2.6.16) and (2.6.17). The inequality (2.6.12) follows by noting that

$$R(s) + D(s)^\top P_\varepsilon(s) D(s) \geqslant 0, \quad \text{a.e. } s \in [0, T] \tag{2.6.18}$$

and passing to the limit in the above. $\qquad\qquad\qquad\qquad\qquad\qquad\qquad\qquad\square$

When the cost functional is not uniformly convex, it is not easy to decide, by means of Theorem 2.3.2, whether Problem (SLQ) is open-loop solvable or not. Even if it is open-loop solvable, finding an open-loop optimal control is not an easy job because the optimality system (2.3.3) is a coupled FBSDE. Fortunately, we are able to construct a minimizing family for Problem (SLQ) when it is finite, as the above theorem (part (i)) showed. As a matter of fact, we have the following result.

Theorem 2.6.2 *Let (H1)–(H2) hold and $(t, x) \in [0, T) \times \mathbb{R}^n$ be a given initial pair. Suppose that (2.6.3) holds, and let u_ε ($\varepsilon > 0$) be defined by (2.6.9). Then the following statements are equivalent:*

(i) *Problem (SLQ) is open-loop solvable at (t, x);*
(ii) *the family $\{u_\varepsilon\}_{\varepsilon>0}$ is bounded in the Hilbert space $\mathcal{U}[t, T]$;*
(iii) *the family $\{u_\varepsilon\}_{\varepsilon>0}$ converges strongly in $\mathcal{U}[t, T]$ as $\varepsilon \to 0$;*
(iv) *the family $\{u_\varepsilon\}_{\varepsilon>0}$ converges weakly in $\mathcal{U}[t, T]$ as $\varepsilon \to 0$.*

Whenever (i), (ii), (iii), or (iv) is satisfied, u_ε converges strongly to an open-loop optimal control of Problem (SLQ) for the initial pair (t, x) as $\varepsilon \to 0$.

Proof The proof follows immediately from Proposition 1.3.4 of Chap. 1.   □

From Theorem 2.6.2, we see that the open-loop solvability of Problem $(SLQ)^0$ at (t, x) is equivalent to the boundedness of $\{\Theta_\varepsilon X_\varepsilon\}_{\varepsilon>0}$ in $\mathcal{U}[t, T]$, where X_ε is the solution to

$$
\begin{cases}
dX_\varepsilon(s) = (A + B\Theta_\varepsilon)X_\varepsilon ds + (C + D\Theta_\varepsilon)X_\varepsilon dW(s), & s \in [t, T], \\
X_\varepsilon(t) = x.
\end{cases}
$$

Since the $L_{\mathbb{F}}^2(\Omega; C([t, T]; \mathbb{R}^n))$-norm of X_ε is dominated by the L^2-norm of Θ_ε, it is expected that the L^2-boundedness of $\{\Theta_\varepsilon\}_{\varepsilon>0}$ will lead to the open-loop solvability of Problem $(SLQ)^0$. In fact, an even stronger result holds.

Proposition 2.6.3 *Let (H1)–(H2) hold and assume (2.6.3). Let $\{\Theta_\varepsilon\}_{\varepsilon>0}$ be the family defined by (2.6.6). If*

$$
\sup_{\varepsilon>0} \int_0^T |\Theta_\varepsilon(s)|^2 ds < \infty, \tag{2.6.19}
$$

then the Riccati equation (2.4.1) is regularly solvable on $[0, T]$. Consequently, Problem $(SLQ)^0$ is closed-loop solvable.

Proof Fix an arbitrary $x \in \mathbb{R}^n$ and let X_ε be the solution to

$$
\begin{cases}
dX_\varepsilon(s) = (A + B\Theta_\varepsilon)X_\varepsilon ds + (C + D\Theta_\varepsilon)X_\varepsilon dW(s), & s \in [0, T], \\
X_\varepsilon(0) = x.
\end{cases}
$$

By Itô's formula, we have for any $t \in [0, T]$,

$$\mathbb{E}|X_\varepsilon(t)|^2 = |x|^2 + \mathbb{E}\int_0^t \Big[|(C + D\Theta_\varepsilon)X_\varepsilon|^2 + 2\langle(A + B\Theta_\varepsilon)X_\varepsilon, X_\varepsilon\rangle\Big]ds$$

$$\leqslant |x|^2 + \int_0^t \Big[|C + D\Theta_\varepsilon|^2 + 2|A + B\Theta_\varepsilon|\Big]\mathbb{E}|X_\varepsilon|^2 ds,$$

and Gronwall's inequality yields

$$\mathbb{E}|X_\varepsilon(t)|^2 \leqslant |x|^2 \exp\left[\int_0^T \Big(|C + D\Theta_\varepsilon|^2 + 2|A + B\Theta_\varepsilon|\Big)ds\right]$$

$$\leqslant |x|^2 \exp\left[K\Big(1 + \int_0^T |\Theta_\varepsilon|^2 ds\Big)\right], \quad \forall t \in [0, T],$$

where $K > 0$ is a constant depending only on A, B, C, and D. It follows that

$$\mathbb{E}\int_0^T |\Theta_\varepsilon X_\varepsilon|^2 ds \leqslant \int_0^T |\Theta_\varepsilon|^2 \mathbb{E}|X_\varepsilon|^2 ds$$

$$\leqslant |x|^2 \exp\left[K\Big(1 + \int_0^T |\Theta_\varepsilon|^2 ds\Big)\right]\int_0^T |\Theta_\varepsilon|^2 ds,$$

which, together with (2.6.19), implies the boundedness of $\{\Theta_\varepsilon X_\varepsilon\}_{\varepsilon>0}$ in $\mathcal{U}[0, T]$. So by Theorem 2.6.2, Problem (SLQ)^0 is open-loop solvable at $(0, x)$, and hence open-loop solvable at time $t = 0$, since x is arbitrary. Then we use Theorem 2.6.1(ii) to conclude that $P(t) \equiv \lim_{\varepsilon\to 0} P_\varepsilon(t)$ exists for all $t \in [0, T]$ and is such that (2.6.12) holds. Now, let $\{\Theta_{\varepsilon_k}\}$ be a weakly convergent subsequence of $\{\Theta_\varepsilon\}$ with weak limit Θ. Since

$$\mathcal{R}(P_\varepsilon) + \varepsilon I \to \mathcal{R}(P) \quad \text{a.e.} \quad \text{as } \varepsilon \to 0$$

and $\{\mathcal{R}(P_\varepsilon) + \varepsilon I\}_{0<\varepsilon\leqslant 1}$ is uniformly bounded, we have

$$\mathcal{S}(P_{\varepsilon_k}) = -[\mathcal{R}(P_{\varepsilon_k}) + \varepsilon_k I]\Theta_{\varepsilon_k} \to -\mathcal{R}(P)\Theta \quad \text{weakly in } L^2.$$

On the other hand, $\mathcal{S}(P_{\varepsilon_k}) \to \mathcal{S}(P)$ strongly in L^2. Thus, we must have

$$-\mathcal{R}(P)\Theta = \mathcal{S}(P), \quad \text{a.e.} \tag{2.6.20}$$

Then it follows from Proposition A.1.5 in Appendix that

$$\mathscr{R}(\mathcal{S}(P)) \subseteq \mathscr{R}(\mathcal{R}(P)), \quad \mathcal{R}(P)^\dagger \mathcal{S}(P) \in \boldsymbol{\Theta}[t, T].$$

It remains to show that P satisfies the Riccati equation (2.4.1). For this we observe that P_{ε_k} has the following integration representation:

$$P_{\varepsilon_k}(t) = G + \int_t^T \Big[\mathcal{Q}(P_{\varepsilon_k}) + \mathcal{S}(P_{\varepsilon_k})^\top \Theta_{\varepsilon_k}\Big]ds.$$

Since $\Theta_{\varepsilon_k} \rightarrow \Theta$ weakly and $\mathcal{S}(P_{\varepsilon_k}) \rightarrow \mathcal{S}(P)$ strongly in L^2, we obtain by letting $k \rightarrow \infty$ and using the dominated convergence theorem that

$$P(t) = G + \int_t^T \Big[\mathcal{Q}(P) + \mathcal{S}(P)^\top \Theta \Big] ds.$$

Differentiating this equation and making use of (2.6.20), we conclude that P solves the Riccati equation (2.4.1).   □

Next, we recall that $M_2(t)$ is given by

$$M_2(t) = \widehat{L}_t^* G \widehat{L}_t + L_t^* Q L_t + S L_t + L_t^* S^\top + R,$$

and that the cost functional $J(t, x; u)$ is convex in u if and only if $M_2(t) \geqslant 0$, which is equivalent to

$$R(s) \geqslant -\Big[\widehat{L}_t^* G \widehat{L}_t + L_t^* Q(s) L_t + S(s) L_t + L_t^* S(s)^\top \Big], \quad s \in [t, T].$$

Notice that the right-hand side of the above is not necessarily positive semi-definite. Thus, unlike the well-known situation for deterministic LQ problems (where $R \geqslant 0$ is necessary for $M_2(t) \geqslant 0$), R does not have to be positive semi-definite. Instead, we have the following result.

Proposition 2.6.4 *Let (H1)–(H2) hold. If (2.6.3) holds, then*

$$R(t) + D(t)^\top M_0(t) D(t) \geqslant 0, \quad \text{a.e. } t \in [0, T]. \tag{2.6.21}$$

Conversely, suppose in addition to (2.6.21) that

$$B = 0, \quad C = 0, \quad S = 0. \tag{2.6.22}$$

Then Problem (SLQ)^0 is closed-loop solvable, and hence (2.6.3) holds.

Proof If (2.6.3) holds, then for any $\varepsilon > 0$, the unique strongly regular solution P_ε to the Riccati equation (2.6.4) satisfies (2.6.16) and (2.6.18). Hence, (2.6.21) holds. Conversely, if in addition to (2.6.21), (2.6.22) holds, then the corresponding Riccati equation becomes

$$\begin{cases} \dot{P}(s) + P(s) A(s) + A(s)^\top P(s) + Q(s) = 0, \quad \text{a.e. } s \in [0, T], \\ P(T) = G, \end{cases}$$

whose solution happens to be M_0. Thanks to (2.6.21), the solution M_0 is easily seen to be regular. Thus, by Theorem 2.4.3, Problem (SLQ)^0 is closed-loop solvable, since in this case the adapted solution to the BSDE (2.4.3) is identically zero, and thereby the condition (ii) in Theorem 2.4.3 automatically holds. Therefore, we have $M_2(0) \geqslant 0$ by Corollary 2.2.4.   □

From (2.6.21), we see that if R is negative definite, then in order for $u \mapsto J(t, x; u)$ to be convex, it is necessary that D is injective, and either G or Q (or both) has to be positive enough to compensate.

We now look at the following case:

$$D = 0, \quad R \gg 0. \tag{2.6.23}$$

Notice that although $D = 0$, our state equation is still an SDE since C is not necessarily zero.

Theorem 2.6.5 *Let (H1)–(H2) and (2.6.23) hold. Then the following statements are equivalent:*

   (i) Problem (SLQ) is finite at $t = 0$.
  (ii) Problem $(SLQ)^0$ is finite at $t = 0$.
 (iii) $M_2(0) \geqslant \delta I$ for some $\delta > 0$.
 (iv) The Riccati equation

$$\begin{cases} \dot{P} + \mathcal{Q}(P) - \mathcal{S}(P)^\top R^{-1} \mathcal{S}(P) = 0, & s \in [0, T], \\ P(T) = G \end{cases} \tag{2.6.24}$$

    admits a unique solution $P \in C([0, T]; \mathbb{S}^n)$.
 (v) Problem (SLQ) is uniquely closed-loop solvable.
 (vi) Problem (SLQ) is uniquely open-loop solvable.

Proof (i) $\Rightarrow$ (ii): By the functional representation (2.2.2), one has

$$\begin{aligned} V(t, x) + V(t, -x) &\leqslant J(t, x; u) + J(t, -x; -u) \\ &= 2[\langle M_2(t)u, u \rangle + 2\langle M_1(t)x, u \rangle + \langle M_0(t)x, x \rangle + c_t] \\ &= 2[J^0(t, x; u) + c_t]. \end{aligned}$$

Hence, if Problem (SLQ) is finite at t, then for any $u \in \mathcal{U}[t, T]$ and $x \in \mathbb{R}^n$,

$$J^0(t, x; u) \geqslant \frac{1}{2}[V(t, x) + V(t, -x)] - c_t > -\infty,$$

which implies that Problem $(SLQ)^0$ is also finite at t.

(ii) $\Rightarrow$ (iii): Suppose that Problem $(SLQ)^0$ is finite at $t = 0$. For $\varepsilon > 0$, let $P_\varepsilon \in C([0, T]; \mathbb{S}^n)$ be the strongly regular solution to the Riccati equation (2.6.4). Theorem 2.6.1(ii) shows that the limit $P(s) \equiv \lim_{\varepsilon \to 0} P_\varepsilon(s)$ exists for all $s \in [0, T]$ and that (2.6.11) holds. Since $D = 0$ and $R \gg 0$, (2.5.9) automatically holds for some $\lambda > 0$. The conclusion then follows immediately from Theorem 2.5.6 and Proposition 2.2.3.

(iii) $\Leftrightarrow$ (iv): In the case of (2.6.23), the Riccati equation (2.4.1) becomes (2.6.24). If $P \in C([0, T]; \mathbb{S}^n)$ is a solution of (2.6.24), then it is automatically strongly regular. Thus, by Theorem 2.5.6, we obtain the equivalence of (iii) and (iv).

The implications (iii) $\Rightarrow$ (v) and (iii) $\Rightarrow$ (vi) were proved in Corollary 2.5.7, The implications (v) $\Rightarrow$ (i) and (vi) $\Rightarrow$ (i) are trivially true. $\square$

Remark 2.6.6 An interesting point of the above is that under the condition (2.6.23), the finiteness of Problem (SLQ) implies the closed-loop solvability of Problem (SLQ).

2.7 An Example

In this section we re-exam Example 2.1.6. We shall show that in this example, the SLQ problem admits a *continuous* open-loop optimal control for every initial pair (t, x). Thus, the problem is open-loop solvable. But, the value function is *not* continuous in t. As we shall see, the associated Riccati equation has a unique solution P, which, however, does *not* satisfy the range condition (ii) of Definition 2.4.1 and therefore is not regular. So this problem is *not* closed-loop solvable.

Example 2.7.1 Recall the SLQ problem of minimizing $J^0(t, x; u) = \mathbb{E}|X(1)|^2$ subject to the one-dimensional state equation

$$\begin{cases} dX(s) = [u_1(s) + u_2(s)]ds + [u_1(s) - u_2(s)]dW(s), \quad s \in [t, 1], \\ X(t) = x. \end{cases}$$

In this example, $u = (u_1, u_2)^\top$ is the control process and

$$A = 0, \quad C = 0, \quad B = (1, 1), \quad D = (1, -1),$$
$$G = 1, \quad Q = 0, \quad S = (0, 0)^\top, \quad R = 0_{2 \times 2}.$$

The associated Riccati equation is

$$\begin{cases} \dot{P} = P^2 (1\ 1) \begin{pmatrix} P & -P \\ -P & P \end{pmatrix}^\dagger \begin{pmatrix} 1 \\ 1 \end{pmatrix} = \frac{P}{4} (1\ 1) \begin{pmatrix} 1 & -1 \\ -1 & 1 \end{pmatrix} \begin{pmatrix} 1 \\ 1 \end{pmatrix} = 0, \\ P(1) = 1, \end{cases}$$

which, obviously, has a unique solution $P \equiv 1$ over $[0, 1]$. For this solution,

$$\mathcal{R}(\mathcal{S}(P)) = \left\{ (a, a)^\top : a \in \mathbb{R} \right\}, \quad \mathcal{R}(\mathcal{R}(P)) = \left\{ (a, -a)^\top : a \in \mathbb{R} \right\}.$$

Thus, the range condition (ii) of Definition 2.4.1 does not hold and hence P is not regular. By Theorem 2.4.3, the problem is not closed-loop solvable.

Now we look at the open-loop solvability. Consider, for each $\varepsilon > 0$, the Riccati equation

$$\begin{cases} \dot{P}_\varepsilon = P_\varepsilon^2 \begin{pmatrix} 1 & 1 \end{pmatrix} \begin{pmatrix} \varepsilon + P_\varepsilon & -P_\varepsilon \\ -P_\varepsilon & \varepsilon + P_\varepsilon \end{pmatrix}^{-1} \begin{pmatrix} 1 \\ 1 \end{pmatrix} = \dfrac{2}{\varepsilon} P_\varepsilon^2, \\ P_\varepsilon(1) = 1. \end{cases}$$

It is easy to verify that its solution is given by

$$P_\varepsilon(s) = \frac{\varepsilon}{\varepsilon + 2 - 2s}, \quad s \in [0, 1].$$

Letting $\varepsilon \to 0$, we obtain

$$P_0(s) \triangleq \lim_{\varepsilon \to 0} P_\varepsilon(s) = \begin{cases} 0, & 0 \leqslant s < 1, \\ 1, & s = 1. \end{cases}$$

Thus, by Theorem 2.6.1(ii), the original problem is finite with value function

$$V^0(t, x) = \begin{cases} 0, & (t, x) \in [0, 1) \times \mathbb{R}, \\ x^2, & (t, x) \in \{1\} \times \mathbb{R}, \end{cases}$$

which is discontinuous at $t = 1$. According to Theorem 2.6.2, in order to find an open-loop optimal control, we need solve the closed-loop system (2.6.8), which, in this example, reads

$$\begin{cases} dX_\varepsilon(s) = (A + B\Theta_\varepsilon)X_\varepsilon ds + (C + D\Theta_\varepsilon)X_\varepsilon dW = -\dfrac{2P_\varepsilon}{\varepsilon} X_\varepsilon ds, \\ X_\varepsilon(t) = x, \end{cases}$$

where

$$\Theta_\varepsilon \triangleq -[\mathcal{R}(P_\varepsilon) + \varepsilon I_2]^{-1} \mathcal{S}(P_\varepsilon) = -\frac{P_\varepsilon}{\varepsilon} \begin{pmatrix} 1 \\ 1 \end{pmatrix} = -\frac{1}{\varepsilon + 2 - 2s} \begin{pmatrix} 1 \\ 1 \end{pmatrix}.$$

By the variation of constants formula,

$$X_\varepsilon(s) = x \exp\left\{ -\int_t^s \frac{2P_\varepsilon(r)}{\varepsilon} dr \right\} = \frac{\varepsilon + 2 - 2s}{\varepsilon + 2 - 2t} x, \quad t \leqslant s \leqslant 1,$$

and hence

$$u_\varepsilon(s) \triangleq \Theta_\varepsilon(s) X_\varepsilon(s) = -\frac{x}{\varepsilon + 2 - 2t} \begin{pmatrix} 1 \\ 1 \end{pmatrix}, \quad t \leqslant s \leqslant 1,$$

which is a constant control. Clearly, for each $t \in [0, 1)$,

$$u_\varepsilon \to -\frac{x}{2 - 2t} \begin{pmatrix} 1 \\ 1 \end{pmatrix} \equiv u_{(t,x)}, \quad \text{in } L^2 \text{ as } \varepsilon \to 0.$$

Thus, by Theorem 2.6.2, the original problem is open-loop solvable at (t, x) and $u_{(t,x)}$ is an open-loop optimal control which is constant valued.

Chapter 3
Linear-Quadratic Optimal Controls in Infinite Horizons

Abstract This chapter is concerned with stochastic linear-quadratic optimal control problems over an infinite horizon. Existence of an admissible control is non-trivial in this case. To tackle this issue, the notion of L^2-stabilizability is introduced. The existence of an admissible control for each initial state turns out to be equivalent to the L^2-stabilizability of the control system, which in turn is equivalent to the existence of a positive solution to an algebraic Riccati equation. Different from finite-horizon problems, the open-loop and closed-loop solvability coincide in the infinite-horizon case, and both can be established by solving for a stabilizing solution to the associated algebraic Riccati equation. As a consequence, every open-loop optimal control admits a closed-loop representation.

Keywords Linear-quadratic · Optimal control · Infinite horizon · L^2-stabilizability · Open-loop solvability · Closed-loop solvability · Algebraic Riccati equation · Stabilizing solution

Recall the setting, together with various spaces introduced at the beginning of Chap. 2. In addition to those, we further introduce the following spaces (with $\mathbb{H}$ being some Euclidean space):

$$L^2_{\mathbb{F}}(\mathbb{H}) = \left\{ \varphi : [0, \infty) \times \Omega \to \mathbb{H} \mid \varphi \text{ is } \mathbb{F}\text{-progressively measurable,} \right.$$
$$\left. \text{and } \mathbb{E} \int_0^\infty |\varphi(t)|^2 dt < \infty \right\},$$

$$\mathcal{X}_{loc}[0, \infty) = \left\{ \varphi : [0, \infty) \times \Omega \to \mathbb{R}^n \mid \varphi \text{ is } \mathbb{F}\text{-adapted, continuous,} \right.$$
$$\left. \text{and } \mathbb{E} \left[\sup_{0 \leqslant t \leqslant T} |\varphi(t)|^2 \right] < \infty \text{ for every } T > 0 \right\},$$

$$\mathcal{X}[0, \infty) = \left\{ \varphi \in \mathcal{X}_{loc}[0, \infty) \mid \mathbb{E} \int_0^\infty |\varphi(t)|^2 dt < \infty \right\}.$$

Clearly, $\mathcal{X}[0, \infty) \subseteq L^2_{\mathbb{F}}(\mathbb{H})$, and by a trivial extension one can also regard $L^2_{\mathbb{F}}(0, T; \mathbb{H})$ as a subspace of $L^2_{\mathbb{F}}(\mathbb{H})$. We take the inner product

$$\langle \varphi, \psi \rangle = \mathbb{E} \int_0^\infty \langle \varphi(t), \psi(t) \rangle dt, \quad \varphi, \psi \in L^2_{\mathbb{F}}(\mathbb{H})$$

so that $L^2_{\mathbb{F}}(\mathbb{H})$ forms a Hilbert space.

3.1 Formulation of the Problem

Consider the following controlled linear SDE on the infinite horizon $[0, \infty)$:

$$\begin{cases} dX(t) = [AX(t) + Bu(t) + b(t)]dt + [CX(t) + Du(t) + \sigma(t)]dW, \\ X(0) = x, \end{cases} \quad (3.1.1)$$

with the quadratic cost functional

$$\begin{aligned} J(x; u) &\triangleq \mathbb{E} \int_0^\infty \Big[\langle QX(t), X(t) \rangle + 2\langle SX(t), u(t) \rangle + \langle Ru(t), u(t) \rangle \\ &\quad + 2\langle q(t), X(t) \rangle + 2\langle \rho(t), u(t) \rangle \Big] dt \\ &= \mathbb{E} \int_0^\infty \Big[\Big\langle \begin{pmatrix} Q & S^\top \\ S & R \end{pmatrix} \begin{pmatrix} X \\ u \end{pmatrix}, \begin{pmatrix} X \\ u \end{pmatrix} \Big\rangle + 2 \Big\langle \begin{pmatrix} q \\ \rho \end{pmatrix}, \begin{pmatrix} X \\ u \end{pmatrix} \Big\rangle \Big] dt, \quad (3.1.2) \end{aligned}$$

where

$$A, C \in \mathbb{R}^{n \times n}, \quad B, D \in \mathbb{R}^{n \times m}, \quad Q \in \mathbb{S}^n, \quad S \in \mathbb{R}^{m \times n}, \quad R \in \mathbb{S}^m$$

are given constant matrices, and

$$b, \sigma, q \in L^2_{\mathbb{F}}(\mathbb{R}^n), \quad \rho \in L^2_{\mathbb{F}}(\mathbb{R}^m)$$

are given processes. In the above, u, which belongs to $L^2_{\mathbb{F}}(\mathbb{R}^m)$, is called the *control process*, $x \in \mathbb{R}^n$ is the *initial state*, and the solution $X(\cdot) \equiv X(\cdot\,; x, u)$ to the SDE (3.1.1) is called the *state process* corresponding to the control u and the initial state x. A control process u is said to be *admissible* with respect to the initial state x if

$$\mathbb{E} \int_0^\infty |X(t; x, u)|^2 dt < \infty.$$

It is easily seen that for any admissible control u with respect to x, $J(x; u)$ is well-defined. We denote the set of admissible controls with respect to x by $\mathcal{U}_{ad}(x)$. The linear-quadratic optimal control problem over an infinite time horizon can now be stated as follows.

Problem (SLQ)$_\infty$. For given initial state $x \in \mathbb{R}^n$, find an admissible control $u^* \in \mathcal{U}_{ad}(x)$ such that

$$J(x; u^*) = \inf_{u \in \mathcal{U}_{ad}(x)} J(x; u) \equiv V(x). \tag{3.1.3}$$

If $u^* \in \mathcal{U}_{ad}(x)$ satisfies (3.1.3), then it is called an *open-loop optimal control* of Problem (SLQ)$_\infty$ for the initial state x, the corresponding state process $X^*(\cdot) \equiv X(\cdot; x, u^*)$ is called an *optimal state process*. The function $V : \mathbb{R}^n \to \mathbb{R}$ is called the *value function* of Problem (SLQ)$_\infty$. In the special case of $b, \sigma, q, \rho = 0$, we denote Problem (SLQ)$_\infty$ by Problem (SLQ)$_\infty^0$, the cost functional by $J^0(x; u)$, and the value function by $V^0(x)$.

3.2 Stability

The first question one encounters with Problem (SLQ)$_\infty$ is the existence of admissible controls. The admissible control set could be empty because for any control process $u \in L_\mathbb{F}^2(\mathbb{R}^m)$, we can only ensure that the corresponding state process X is locally square-integrable:

$$\mathbb{E} \int_0^T |X(t)|^2 dt < \infty, \quad \forall 0 < T < \infty.$$

Example 3.2.1 Consider the one-dimensional controlled system

$$dX(t) = X(t)dt + u(t)dW(t), \quad t \geqslant 0.$$

We claim that the admissible control set $\mathcal{U}_{ad}(x)$ is empty for any initial state $x \neq 0$. Indeed, if $u \in L_\mathbb{F}^2(\mathbb{R})$ is admissible, then by definition, the corresponding controlled state process X satisfies

$$\int_0^\infty \mathbb{E}|X(t)|^2 dt = \mathbb{E} \int_0^\infty |X(t)|^2 dt < \infty.$$

This implies $\mathbb{E}|X(t)|^2$, and hence $\mathbb{E}[X(t)]$, goes to zero as $t \to \infty$. Now taking expectation we see $\mathbb{E}[X(t)]$ solves the following ODE:

$$d\mathbb{E}[X(t)] = \mathbb{E}[X(t)]dt; \quad X(0) = x.$$

By the variation of constants formula, we have $\mathbb{E}[X(t)] = xe^t$, which does not have finite limit when $x \neq 0$. This justifies our claim.

In order to settle the question of existence of admissible controls, we shall introduce the concept of *stability* for stochastic linear systems. Consider the following uncontrolled linear system:

$$dX(t) = AX(t)dt + CX(t)dW(t), \quad t \geqslant 0,$$

which we briefly denote by $[A, C]$.

Definition 3.2.2 System $[A, C]$ is said to be L^2-*stable* if for any initial state $x \in \mathbb{R}^n$, its solution $X(\cdot\,; x)$ belongs to the space $\mathcal{X}[0, \infty)$, that is,

$$\mathbb{E} \int_0^\infty |X(t; x)|^2 dt < \infty, \quad \forall x \in \mathbb{R}^n.$$

Let $\Phi = \{\Phi(t); t \geqslant 0\}$ be the solution to the matrix SDE

$$\begin{cases} d\Phi(t) = A\Phi(t)dt + C\Phi(t)dW(t), \quad t \geqslant 0, \\ \Phi(0) = I_n. \end{cases} \tag{3.2.1}$$

The following result provides a characterization of the L^2-stability of $[A, C]$.

Theorem 3.2.3 *The system $[A, C]$ is L^2-stable if and only if there exists a $P \in \mathbb{S}_+^n$ such that*

$$PA + A^\top P + C^\top PC < 0. \tag{3.2.2}$$

In this case, for any $\Lambda \in \mathbb{S}^n$, the Lyapunov equation

$$PA + A^\top P + C^\top PC + \Lambda = 0$$

admits a unique solution $P \in \mathbb{S}^n$ given by

$$P = \mathbb{E} \int_0^\infty \Phi(t)^\top \Lambda \Phi(t)dt,$$

where Φ is the solution of (3.2.1).

Proof *Sufficiency.* For any fixed $\Lambda \in \mathbb{S}^n$, consider the linear ODE on $[0, \infty)$:

$$\dot{\Theta}(t) = \Theta(t)A + A^\top \Theta(t) + C^\top \Theta(t)C + \Lambda; \quad \Theta(0) = 0. \tag{3.2.3}$$

Clearly, it has a unique solution $\Theta(t)$ defined on $[0, \infty)$, and for any fixed $\tau > 0$, the function

$$\Theta_\tau(s) = \Theta(\tau - s), \quad s \in [0, \tau]$$

solves the equation

$$\dot{\Theta}_\tau(s) + \Theta_\tau(s)A + A^\top \Theta_\tau(s) + C^\top \Theta_\tau(s)C + \Lambda = 0; \quad \Theta_\tau(\tau) = 0$$

on the interval $[0, \tau]$. Let $X(\cdot) \equiv X(\cdot\,; x)$ be the solution to system $[A, C]$ with initial state x and note that $X(s)$ has the representation $X(s) = \Phi(s)x$. Applying Itô's formula to $s \mapsto \langle \Theta_\tau(s)X(s), X(s)\rangle$, we obtain

$$
\begin{aligned}
-\langle \Theta_\tau(0)x, x\rangle &= \mathbb{E}\big[\langle \Theta_\tau(\tau)X(\tau), X(\tau)\rangle - \langle \Theta_\tau(0)x, x\rangle\big] \\
&= \mathbb{E}\int_0^\tau \langle (\dot{\Theta}_\tau + \Theta_\tau A + A^\top \Theta_\tau + C^\top \Theta_\tau C)X, X\rangle(s)ds \\
&= -\mathbb{E}\int_0^\tau \langle \Lambda X(s), X(s)\rangle ds \\
&= -x^\top\left[\mathbb{E}\int_0^\tau \Phi(s)^\top \Lambda \Phi(s)ds\right]x.
\end{aligned}
$$

It follows that

$$\Theta(\tau) = \Theta_\tau(0) = \mathbb{E}\int_0^\tau \Phi(s)^\top \Lambda \Phi(s)ds, \quad \tau \geqslant 0.$$

If the system $[A, C]$ is L^2-stable, one has the following limit:

$$\lim_{\tau \to \infty} \Theta(\tau) = \mathbb{E}\int_0^\infty \Phi(s)^\top \Lambda \Phi(s)ds \equiv P.$$

Because $\Theta(t)$ is the solution to (3.2.3), we have for any $t > 0$,

$$
\begin{aligned}
\Theta(t+1) - \Theta(t) &= \left(\int_t^{t+1} \Theta(s)ds\right)A + A^\top\left(\int_t^{t+1} \Theta(s)ds\right) \\
&\quad + C^\top\left(\int_t^{t+1} \Theta(s)ds\right)C + \Lambda.
\end{aligned}
$$

Letting $t \to \infty$, we obtain

$$PA + A^\top P + C^\top PC + \Lambda = 0.$$

In particular, if we take $\Lambda = I_n$, then the corresponding P is positive definite and satisfies (3.2.2).

Necessity. Suppose $P \in \mathbb{S}_+^n$ satisfies (3.2.2). Let $X(\cdot) \equiv X(\cdot\,; x)$ be the solution to system $[A, C]$ with initial state x. By Itô's formula, we have for any $t > 0$,

$$\mathbb{E}\langle PX(t), X(t)\rangle - \langle Px, x\rangle = \mathbb{E}\int_0^t \langle (PA + A^\top P + C^\top PC)X(s), X(s)\rangle ds.$$

Let $\lambda > 0$ be the smallest eigenvalue of $-(PA + A^\top P + C^\top PC)$. Then

$$\lambda \mathbb{E} \int_0^t |X(s)|^2 ds \leqslant -\mathbb{E} \int_0^t \langle (PA + A^\top P + C^\top PC)X(s), X(s) \rangle ds$$
$$\leqslant \langle Px, x \rangle - \mathbb{E} \langle PX(t), X(t) \rangle$$
$$\leqslant \langle Px, x \rangle, \quad \forall t > 0,$$

which implies the L^2-stability of $[A, C]$.   □

Now let us look at the nonhomogeneous system

$$dX(t) = [AX(t) + \varphi(t)]dt + [CX(t) + \rho(t)]dW(t), \quad t \geqslant 0. \qquad (3.2.4)$$

Proposition 3.2.4 *Suppose that $[A, C]$ is L^2-stable. Then for any $\varphi, \rho \in L^2_{\mathbb{F}}(\mathbb{R}^n)$ and any initial state $x \in \mathbb{R}^n$, the solution $X(\cdot) \equiv X(\cdot\,; x, \varphi, \rho)$ of (3.2.4) is in $\mathcal{X}[0, \infty)$. Moreover, there exists a constant $K > 0$, independent of x, φ and ρ, such that*

$$\mathbb{E} \int_0^\infty |X(t)|^2 dt \leqslant K \left\{ |x|^2 + \mathbb{E} \int_0^\infty \left[|\varphi(t)|^2 + |\rho(t)|^2 \right] dt \right\}.$$

Proof Since $[A, C]$ is L^2-stable, by Theorem 3.2.3, there exists a $P > 0$ such that

$$PA + A^\top P + C^\top PC + I_n = 0.$$

Applying Itô's formula to $s \mapsto \langle PX(s), X(s) \rangle$, we obtain for all $t > 0$,

$$\mathbb{E} \langle PX(t), X(t) \rangle - \langle Px, x \rangle$$
$$= \mathbb{E} \int_0^t \Big[\langle (PA + A^\top P + C^\top PC)X(s), X(s) \rangle$$
$$+ 2\langle P\varphi(s) + C^\top P\rho(s), X(s) \rangle + \langle P\rho(s), \rho(s) \rangle \Big] ds$$
$$= \mathbb{E} \int_0^t \Big[-|X(s)|^2 + 2\langle P\varphi(s) + C^\top P\rho(s), X(s) \rangle + \langle P\rho(s), \rho(s) \rangle \Big] ds.$$

Let $\lambda > 0$ be the smallest eigenvalue of P and set

$$\alpha(s) = P\varphi(s) + C^\top P\rho(s), \quad \beta(s) = \langle P\rho(s), \rho(s) \rangle; \quad s > 0.$$

Then by the Cauchy-Schwarz inequality,

$$\lambda \mathbb{E}|X(t)|^2 \leqslant \mathbb{E}\langle PX(t), X(t)\rangle$$

$$\leqslant \langle Px, x\rangle + \mathbb{E}\int_0^t \left[-|X(s)|^2 + 2\langle \alpha(s), X(s)\rangle + \beta(s)\right]ds$$

$$\leqslant \langle Px, x\rangle + \mathbb{E}\int_0^t \left[-\frac{1}{2}|X(s)|^2 + 2|\alpha(s)|^2 + \beta(s)\right]ds$$

$$= \langle Px, x\rangle + \int_0^t \left[-\frac{1}{2}\mathbb{E}|X(s)|^2 + 2\mathbb{E}|\alpha(s)|^2 + \mathbb{E}\beta(s)\right]ds.$$

It follows from Gronwall's inequality that

$$\lambda \mathbb{E}|X(t)|^2 \leqslant \langle Px, x\rangle e^{-(2\lambda)^{-1}t} + \int_0^t e^{-(2\lambda)^{-1}(t-s)}\left[2\mathbb{E}|\alpha(s)|^2 + \mathbb{E}\beta(s)\right]ds,$$

which, together with Young's inequality, implies the integrability of $\mathbb{E}|X(t)|^2$ over $[0, \infty)$. $\qquad\square$

3.3 Stabilizability

According to Proposition 3.2.4, when the system $[A, C]$ is L^2-stable, the admissible control set $\mathcal{U}_{ad}(x)$ is nonempty (actually equals $L^2_{\mathbb{F}}(\mathbb{R}^n)$) for all $x \in \mathbb{R}^n$. The following simple example shows that the converse is not true.

Example 3.3.1 Consider the one-dimensional controlled system

$$dX(t) = [X(t) + u(t)]dt + [X(t) + u(t)]dW(t), \quad t \geqslant 0.$$

In this example, $A = C = 1$. It is clear, by Theorem 3.2.3, that the system $[A, C]$ is not L^2-stable. However, for any initial state $x \in \mathbb{R}$, the control defined by

$$\hat{u}(t) = -2xe^{-W(t)-\frac{3}{2}t}, \quad t > 0$$

is admissible with respect to x. In fact, one can verify, using Itô's formula, that the solution to

$$\begin{cases} dX(t) = [X(t) + \hat{u}(t)]dt + [X(t) + \hat{u}(t)]dW(t), \quad t \geqslant 0, \\ X(0) = x, \end{cases}$$

is given by

$$X(t) = xe^{-W(t)-\frac{3}{2}t}.$$

Noting that $e^{-2W(t)-2t}$ is a martingale, we have

$$\mathbb{E} \int_0^\infty |X(t)|^2 dt = \int_0^\infty x^2 e^{-t} \mathbb{E}\big[e^{-2W(t)-2t}\big] dt = x^2 \int_0^\infty e^{-t} dt < \infty.$$

Thus, $\hat{u}$ is admissible with respect to x.

3.3.1 Definition and Characterization

In order to characterize the admissible control sets, we further introduce the concept of *stabilizability*. Denote by $[A, C; B, D]$ the following controlled linear system:

$$dX(t) = [AX(t) + Bu(t)]dt + [CX(t) + Du(t)]dW(t), \quad t \geqslant 0.$$

Definition 3.3.2 System $[A, C; B, D]$ is said to be L^2-*stabilizable* if there exists a matrix $\Theta \in \mathbb{R}^{m \times n}$ such that $[A + B\Theta, C + D\Theta]$ is L^2-stable. In this case, Θ is called a *stabilizer* of $[A, C; B, D]$. The set of all stabilizers of $[A, C; B, D]$ is denoted by $\mathscr{S} \equiv \mathscr{S}[A, C; B, D]$.

The following result shows that the L^2-stabilizability is sufficient for the existence of an admissible control and gives an explicit description of the admissible control sets.

Proposition 3.3.3 *Suppose that* $\Theta \in \mathscr{S}[A, C; B, D]$. *Then for any* $x \in \mathbb{R}^n$,

$$\mathcal{U}_{ad}(x) = \big\{\Theta X_\Theta(\cdot; x, v) + v : v \in L^2_{\mathbb{F}}(\mathbb{R}^m)\big\},$$

where $X_\Theta(\cdot; x, v)$ *is the solution to the SDE*

$$\begin{cases} dX_\Theta(t) = [(A + B\Theta)X_\Theta(t) + Bv(t) + b(t)]dt \\ \qquad\qquad + [(C + D\Theta)X_\Theta(t) + Dv(t) + \sigma(t)]dW(t), \quad t \geqslant 0, \qquad (3.3.1) \\ X_\Theta(0) = x. \end{cases}$$

Proof Let $v \in L^2_{\mathbb{F}}(\mathbb{R}^m)$ and let $X_\Theta(\cdot) \equiv X_\Theta(\cdot; x, v)$ be the corresponding solution to (3.3.1). Since $[A + B\Theta, C + D\Theta]$ is L^2-stable, by Proposition 3.2.4, $X_\Theta \in \mathcal{X}[0, \infty)$. Set

$$u = \Theta X_\Theta + v \in L^2_{\mathbb{F}}(\mathbb{R}^m),$$

and let $X \in \mathcal{X}_{loc}[0, \infty)$ be the solution to

$$\begin{cases} dX(t) = [AX(t) + Bu(t) + b(t)]dt \\ \qquad\qquad + [CX(t) + Du(t) + \sigma(t)]dW(t), \quad t \geqslant 0, \qquad (3.3.2) \\ X(0) = x. \end{cases}$$

By the uniqueness of solutions, $X = X_\Theta \in \mathcal{X}[0, \infty)$ and hence $u \in \mathcal{U}_{ad}(x)$.

On the other hand, suppose that $u \in \mathcal{U}_{ad}(x)$ and let $X \in \mathcal{X}[0, \infty)$ be the corresponding solution of (3.3.2). Then with the control v defined by

$$v \triangleq u - \Theta X \in L_{\mathbb{F}}^2(\mathbb{R}^m),$$

the solution X_Θ of (3.3.1) coincides with X, again by the uniqueness of solutions. Thus, u admits a representation of the form $\Theta X_\Theta(\cdot\,; x, v) + v$. $\square$

We now provide a characterization for non-emptiness of admissible control sets in terms of the L^2-stabilizability. The result further shows that the L^2-stabilizability is not only sufficient, but also necessary, for the non-emptiness of $\mathcal{U}_{ad}(x)$ for all $x \in \mathbb{R}^n$. Recall the notation

$$\begin{cases} \mathcal{Q}(P) = PA + A^\top P + C^\top PC + Q, \\ \mathcal{S}(P) = B^\top P + D^\top PC + S, \\ \mathcal{R}(P) = R + D^\top PD, \end{cases} \tag{3.3.3}$$

for $P \in \mathbb{S}^n$. We first present the following lemma.

Lemma 3.3.4 *Suppose that for each $T > 0$ the differential Riccati equation*

$$\begin{cases} \dot{P}_\tau(s) + \mathcal{Q}(P_\tau(s)) - \mathcal{S}(P_\tau(s))^\top \mathcal{R}(P_\tau(s))^{-1} \mathcal{S}(P_\tau(s)) = 0, \\ P_\tau(T) = G \end{cases} \tag{3.3.4}$$

admits a solution $P_\tau \in C([0, T]; \mathbb{S}^n)$ such that

$$\mathcal{R}(P_\tau(s)) > 0, \quad \forall s \in [0, T].$$

If $P_\tau(0)$ converges to P as $T \to \infty$ and $\mathcal{R}(P)$ is invertible, then P solves the following algebraic Riccati equation:

$$\mathcal{Q}(P) - \mathcal{S}(P)^\top \mathcal{R}(P)^{-1} \mathcal{S}(P) = 0. \tag{3.3.5}$$

Proof For fixed but arbitrary $0 < T_1 < T_2 < \infty$, we define

$$\begin{cases} P_1(s) = P_{T_1}(T_1 - s), \quad 0 \leqslant s \leqslant T_1, \\ P_2(s) = P_{T_2}(T_2 - s), \quad 0 \leqslant s \leqslant T_2, \end{cases}$$

and denote

$$\Theta_i(s) = \mathcal{R}(P_i(s))^{-1} \mathcal{S}(P_i(s)), \quad i = 1, 2.$$

Note that on the interval $[0, T_1]$, both P_1 and P_2 solve the same equation

$$\begin{cases} \dot{\Sigma}(s) - \mathcal{Q}(\Sigma(s)) + \mathcal{S}(\Sigma(s))\mathcal{R}(\Sigma(s))^{-1}\mathcal{S}(\Sigma(s)) = 0, \\ \Sigma(0) = G. \end{cases} \tag{3.3.6}$$

Hence, by the uniqueness of solutions to ODEs,

$$P_1(s) = P_2(s), \quad \forall s \in [0, T_1].$$

In fact, the difference $\Delta = P_1 - P_2$ satisfies $\Delta(0) = 0$ and

$$\dot{\Delta} = \Delta A + A^\top \Delta + C^\top \Delta C - (\Delta B + C^\top \Delta D)\Theta_1 \\ + \Theta_2^\top D^\top \Delta D \Theta_1 - \Theta_2^\top (B^\top \Delta + C^\top \Delta D), \quad s \in [0, T_1].$$

By assumption, Θ_1 and Θ_2 are continuous and hence bounded. Thus, we have for some constant $K > 0$ independent of Δ,

$$\begin{aligned} |\Delta(t)| &\leqslant \int_0^t \Big| \Delta A + A^\top \Delta + C^\top \Delta C - (\Delta B + C^\top \Delta D)\Theta_1 \\ &\qquad + \Theta_2^\top D^\top \Delta D \Theta_1 - \Theta_2^\top (B^\top \Delta + C^\top \Delta D) \Big| ds \\ &\leqslant K \int_0^t |\Delta(s)| ds, \quad \forall t \in [0, T_1]. \end{aligned}$$

It follows from Gronwall's inequality that $\Delta(s) = 0$ for all $s \in [0, T_1]$. Therefore, we may define a function $\Sigma : [0, \infty) \to \mathbb{S}^n$ by the following:

$$\Sigma(s) = P_T(T - s), \quad \text{if } 0 \leqslant s \leqslant T.$$

If $\Sigma(T) = P_T(0)$ converges to P as $T \to \infty$ and $\mathcal{R}(P)$ is invertible, then with

$$\begin{aligned} \Pi(s) &\triangleq \mathcal{Q}(\Sigma(s)) - \mathcal{S}(\Sigma(s))^\top \mathcal{R}(\Sigma(s))^{-1}\mathcal{S}(\Sigma(s)), \\ \Pi_\infty &\triangleq \mathcal{Q}(P) - \mathcal{S}(P)^\top \mathcal{R}(P)^{-1}\mathcal{S}(P), \end{aligned}$$

we have $\lim_{s \to \infty} \Pi(s) = \Pi_\infty$. On the other hand, since Σ satisfies (3.3.6) on the whole interval $[0, \infty)$, we have

$$\Sigma(T + 1) - \Sigma(T) = \int_T^{T+1} \Pi(t)dt, \quad \forall T > 0.$$

It follows that

$$|\Pi_\infty| \leqslant \left| \int_T^{T+1} \Pi(t)dt \right| + \left| \int_T^{T+1} \left[\Pi_\infty - \Pi(t) \right] dt \right|$$

$$\leqslant |\Sigma(T+1) - \Sigma(T)| + \int_T^{T+1} |\Pi_\infty - \Pi(t)| dt.$$

The desired result then follows by letting $T \to \infty$ in the above. □

Theorem 3.3.5 *The following statements are equivalent:*

(i) $\mathcal{U}_{ad}(x) \neq \varnothing$ *for all* $x \in \mathbb{R}^n$;

(ii) $\mathscr{S}[A, C; B, D] \neq \varnothing$;

(iii) *The following algebraic Riccati equation (ARE, for short) admits a positive solution* $P \in \mathbb{S}_+^n$:

$$PA + A^\top P + C^\top PC + I \tag{3.3.7}$$
$$- (PB + C^\top PD)(I + D^\top PD)^{-1}(B^\top P + D^\top PC) = 0.$$

If the above are satisfied and P is a positive solution of (3.3.7), then

$$\Gamma \triangleq -(I + D^\top PD)^{-1}(B^\top P + D^\top PC) \in \mathscr{S}[A, C; B, D]. \tag{3.3.8}$$

Proof We have proved the implication (ii) ⇒ (i) in Proposition 3.3.3. For the implication (iii) ⇒ (ii), we observe that if P is a positive definite solution of (3.3.7) and Γ is defined by (3.3.8), then

$$P(A + B\Gamma) + (A + B\Gamma)^\top P + (C + D\Gamma)^\top P(C + D\Gamma) = -I - \Gamma^\top \Gamma < 0.$$

Hence, by Theorem 3.2.3 and Definition 3.3.2, Γ is stabilizer of $[A, C; B, D]$.

We next show that (i) ⇒ (iii). By subtracting solutions of the state equation (3.1.1) corresponding to x and 0, we may assume without loss of generality that $b = \sigma = 0$. Let $e_1, \ldots, e_n$ be the standard basis for $\mathbb{R}^n$. Take $u_i \in \mathcal{U}_{ad}(e_i), i = 1, \ldots, n$, and set $U = (u_1, \ldots, u_n)$. Then, by the linearity of the state equation, $Ux \in \mathcal{U}_{ad}(x)$ for all $x \in \mathbb{R}^n$. Consider the cost functional

$$\bar{J}(x; u) = \mathbb{E} \int_0^\infty \left[|X(t)|^2 + |u(t)|^2 \right] dt.$$

With $\mathbb{X} \in L_{\mathbb{F}}^2(\mathbb{R}^{n \times n})$ being the solution to the matrix SDE

$$\begin{cases} d\mathbb{X}(t) = [A\mathbb{X}(t) + BU(t)]dt + [C\mathbb{X}(t) + DU(t)]dW(t), \quad t \geqslant 0, \\ \mathbb{X}(0) = I_n, \end{cases}$$

we have for any $x \in \mathbb{R}^n$,

$$
\begin{aligned}
\inf_{u \in \mathcal{U}_{ad}(x)} \bar{J}(x; u) &\leqslant \mathbb{E} \int_0^\infty \Big[|\mathbb{X}(t)x|^2 + |U(t)x|^2 \Big] dt \\
&= \left\langle \left(\mathbb{E} \int_0^\infty \Big[\mathbb{X}(t)^\top \mathbb{X}(t) + U(t)^\top U(t) \Big] dt \right) x, x \right\rangle.
\end{aligned}
\tag{3.3.9}
$$

Now for a fixed but arbitrary $T > 0$, let us consider the optimal control problem in the finite time horizon $[0, T]$ with state equation

$$
\begin{cases}
dX_T(t) = [AX_T(t) + Bu(t)]dt + [CX_T(t) + Du(t)]dW(t), & t \in [0, T], \\
X_T(0) = x,
\end{cases}
$$

and cost functional

$$
\bar{J}_T(x; u) = \mathbb{E} \int_0^T \Big[|X_T(t)|^2 + |u(t)|^2 \Big] dt.
$$

By Theorem 2.5.6 of Chap. 2, the differential Riccati equation

$$
\begin{cases}
\dot{P}_T(t) + P_T(t)A + A^\top P_T(t) + C^\top P_T(t)C + I \\
\quad - \big[P_T(t)B + C^\top P_T(t)D \big] \big[I + D^\top P_T(t)D \big]^{-1} \\
\quad \times \big[B^\top P_T(t) + D^\top P_T(t)C \big] = 0, \quad t \in [0, T], \\
P_T(T) = 0
\end{cases}
$$

admits a unique solution $P_T \in C([0, T]; \mathbb{S}_+^n)$ such that

$$
\langle P_T(0)x, x \rangle = \inf_{u \in L_{\mathbb{F}}^2(0, T; \mathbb{R}^m)} \bar{J}_T(x; u), \quad \forall x \in \mathbb{R}^n.
$$

Since for any $u \in \mathcal{U}_{ad}(x)$, the restriction $u|_{[0,T]}$ of u to $[0, T]$ belongs to $\mathcal{U}[0, T] \equiv L_{\mathbb{F}}^2(0, T; \mathbb{R}^m)$, we have

$$
\langle P_T(0)x, x \rangle \leqslant \bar{J}_T(x; u|_{[0,T]}) \leqslant \bar{J}(x; u), \quad \forall u \in \mathcal{U}_{ad}(x),
$$

which, together with (3.3.9), implies

$$
\langle P_T(0)x, x \rangle \leqslant \langle \Lambda x, x \rangle, \quad \forall x \in \mathbb{R}^n,
\tag{3.3.10}
$$

where $\langle \Lambda x, x \rangle$ denotes the right-hand side of (3.3.9). On the other hand, for any fixed $T' > T > 0$, the restriction $u|_{[0,T]}$ of $u \in L_{\mathbb{F}}^2(0, T'; \mathbb{R}^m)$ also belongs to $L_{\mathbb{F}}^2(0, T; \mathbb{R}^m)$. Thus,

$$
\langle P_T(0)x, x \rangle \leqslant \bar{J}_T(x; u|_{[0,T]}) \leqslant \bar{J}_{T'}(x; u), \quad \forall u \in L_{\mathbb{F}}^2(0, T'; \mathbb{R}^m),
$$

which in turn gives

$$\langle P_T(0)x, x \rangle \leqslant \langle P_{T'}(0)x, x \rangle, \quad \forall x \in \mathbb{R}^n. \tag{3.3.11}$$

Combining (3.3.10)–(3.3.11) and noting that $P_T \in C([0, T]; \mathbb{S}^n_+)$, we obtain

$$0 < P_T(0) \leqslant P_{T'}(0) \leqslant \Lambda, \quad \forall 0 < T < T' < \infty.$$

This implies that $P_T(0)$ converges increasingly to some $P \in \mathbb{S}^n_+$ as $T \nearrow \infty$. By Lemma 3.3.4, the limit matrix P solves the ARE (3.3.7).   □

3.3.2   The Case of One-Dimensional State

In this subsection, we look at the case $n = 1$, i.e., the state variable is one-dimensional. However, the control is still allowed to be multi-dimensional.

Lemma 3.3.6 *Let* $n = 1$. *If system* $[A, C; B, D]$ *is not* L^2*-stabilizable, then*

$$\begin{pmatrix} 2A + C^2 & B + CD \\ B^\top + CD^\top & D^\top D \end{pmatrix} \geqslant 0. \tag{3.3.12}$$

Proof If $[A, C; B, D]$ is not L^2-stabilizable, then by Definition 3.3.2 and Theorem 3.2.3, we have

$$2(A + B\Theta) + (C + D\Theta)^2 \geqslant 0, \quad \forall \Theta \in \mathbb{R}^m.$$

Since for any $x \in \mathbb{R} \setminus \{0\}$ and any $y \in \mathbb{R}^m$ one can find a $\Theta \in \mathbb{R}^m$ such that $y = \Theta x$, we have

$$\begin{aligned}
(x \ y^\top) &\begin{pmatrix} 2A + C^2 & B + CD \\ B^\top + CD^\top & D^\top D \end{pmatrix} \begin{pmatrix} x \\ y \end{pmatrix} \\
&= x \left(1 \ \Theta^\top\right) \begin{pmatrix} 2A + C^2 & B + CD \\ B^\top + CD^\top & D^\top D \end{pmatrix} \begin{pmatrix} 1 \\ \Theta \end{pmatrix} x \\
&= \left[2(A + B\Theta) + (C + D\Theta)^2\right] x^2 \geqslant 0,
\end{aligned}$$

for all $x \neq 0$ and all $y \in \mathbb{R}^m$. The result follows immediately.   □

For the moment, let us assume that $b = \sigma = 0$ and denote

$$\mathcal{V} = \left\{ u \in L^2_{\mathbb{F}}(\mathbb{R}^m) \mid Bu = Du = 0, \text{ a.e. a.s.} \right\}.$$

Obviously, $0 \in \mathcal{V} \subseteq \mathcal{U}_{ad}(0)$, and hence $\mathcal{U}_{ad}(0)$ is non-empty.

Theorem 3.3.7 *Let* $n = 1$, *and suppose that* $b = \sigma = 0$. *Then exactly one of the following holds:*

(i) $\mathcal{U}_{ad}(0) = \mathcal{V}$ *and* $\mathcal{U}_{ad}(x) = \varnothing$ *for all* $x \neq 0$.
(ii) *The system* $[A, C; B, D]$ *is* L^2*-stabilizable.*

Proof We prove it by contradiction. From Theorem 3.3.5 we see that (i) and (ii) cannot hold simultaneously. Now suppose that neither (i) nor (ii) holds. Then either $\mathcal{U}_{ad}(0)\backslash\mathcal{V} \neq \varnothing$ or else $\mathcal{U}_{ad}(x) \neq \varnothing$ for some $x \neq 0$, and (3.3.12) holds by Lemma 3.3.6. If there exists a $u \in \mathcal{U}_{ad}(0)\backslash\mathcal{V}$, then with X_0 denoting the solution of (3.1.1) corresponding to the initial state $x = 0$ and the admissible control u, we have

$$\mathbb{E}|X_0(t)|^2 = \mathbb{E}\int_0^t \left\{2[AX_0(s) + Bu(s)]X_0(s) + |CX_0(s) + Du(s)|^2\right\}ds$$
$$= \mathbb{E}\int_0^t \left\langle \begin{pmatrix} 2A + C^2 & B + CD \\ B^\top + CD^\top & D^\top D \end{pmatrix}\begin{pmatrix} X_0(s) \\ u(s) \end{pmatrix}, \begin{pmatrix} X_0(s) \\ u(s) \end{pmatrix}\right\rangle ds \geq 0.$$

Since (3.3.12) holds and $X_0 \in \mathcal{X}[0, \infty)$ (and hence $\lim_{t\to\infty}\mathbb{E}|X_0(t)|^2 = 0$), the integrand in the above must vanish for all $s \geq 0$. It turns out that $X_0(s) = 0$ for all $s \geq 0$, and hence
$$Bu = 0, \quad Du = 0, \quad \text{a.e. a.s.,}$$

which means $u \in \mathcal{V}$, a contradiction. Now if $\mathcal{U}_{ad}(x) \neq \varnothing$ for some $x \neq 0$, take $v \in \mathcal{U}_{ad}(x)$ and let X be the solution of (3.1.1) corresponding to x and v. Then, using (3.3.12), we have for any $t \geq 0$,

$$\mathbb{E}|X(t)|^2 - |x|^2 = \mathbb{E}\int_0^t \left\langle \begin{pmatrix} 2A + C^2 & B + CD \\ B^\top + CD^\top & D^\top D \end{pmatrix}\begin{pmatrix} X(s) \\ v(s) \end{pmatrix}, \begin{pmatrix} X(s) \\ v(s) \end{pmatrix}\right\rangle ds \geq 0,$$

which is impossible since $\lim_{t\to\infty}\mathbb{E}|X(t)|^2 = 0$. This completes the proof. □

For the case $b \neq 0$ or $\sigma \neq 0$, we have the following result.

Theorem 3.3.8 *Let* $n = 1$, *and suppose that* $b \neq 0$ *or* $\sigma \neq 0$. *Then exactly one of the following holds:*

(i) $\mathcal{U}_{ad}(x) = \varnothing$ *for all* $x \in \mathbb{R}^n$.
(ii) *There is only one* $x \in \mathbb{R}^n$ *for which the admissible control set* $\mathcal{U}_{ad}(x) \neq \varnothing$. *In this case,*
$$u - v \in \mathcal{V}, \quad \forall u, v \in \mathcal{U}_{ad}(x).$$

(iii) *The system* $[A, C; B, D]$ *is* L^2*-stabilizable, or equivalently,* $\mathcal{U}_{ad}(x) \neq \varnothing$ *for all* $x \in \mathbb{R}^n$.

Proof Clearly, any two of the statements (i)–(iii) cannot hold simultaneously. Now let us assume that neither (i) nor (ii) holds. Then

$$\mathcal{U}_{ad}(x_1) \neq \varnothing, \quad \mathcal{U}_{ad}(x_2) \neq \varnothing$$

for some $x_1 \neq x_2$. Take $u_i \in \mathcal{U}_{ad}(x_i)$, $i = 1, 2$, and let X_i be the solution of (3.1.1) corresponding to the initial state x_i and the admissible control u_i. Then with $x = x_1 - x_2$ and $u = u_1 - u_2$, the process $X \triangleq X_1 - X_2$ is in $\mathcal{X}[0, \infty)$ and solves

$$\begin{cases} dX(t) = [AX(t) + Bu(t)]dt + [CX(t) + Du(t)]dW(t), & t \geqslant 0, \\ X(0) = x. \end{cases}$$

Thus, by Theorem 3.3.7, the system $[A, C; B, D]$ is L^2-stabilizable.

Suppose that there is only one $x \in \mathbb{R}^n$ such that $\mathcal{U}_{ad}(x) \neq \varnothing$. The same argument as before shows that for any $u, v \in \mathcal{U}_{ad}(x)$, the solution X_0 of

$$\begin{cases} dX_0(t) = \big\{AX_0(t) + B[u(t) - v(t)]\big\}dt + \big\{CX_0(t) + D[u(t) - v(t)]\big\}dW, \\ X_0(0) = 0, \end{cases}$$

is in $\mathcal{X}[0, \infty)$. Since $\mathscr{S}[A, C; B, D] = \varnothing$ in this situation, we have $u - v \in \mathcal{V}$ by Theorem 3.3.7. $\qquad\square$

3.4 Solvability and the Algebraic Riccati Equation

Let us return to Problem $(SLQ)_\infty$. According to Theorem 3.3.5, Problem $(SLQ)_\infty$ is well-posed (for all $x \in \mathbb{R}^n$) only if the system $[A, C; B, D]$ is L^2-stabilizable. Therefore, it is reasonable to assume the following:

(S) System $[A, C; B, D]$ is L^2-stabilizable, i.e., $\mathscr{S}[A, C; B, D] \neq \varnothing$.

Definition 3.4.1 An element $u^* \in \mathcal{U}_{ad}(x)$ is called an *open-loop optimal control* of Problem $(SLQ)_\infty$ for the initial state $x \in \mathbb{R}^n$ if

$$J(x; u^*) \leqslant J(x; u), \quad \forall u \in \mathcal{U}_{ad}(x).$$

If an open-loop optimal control (uniquely) exists for x, Problem $(SLQ)_\infty$ is said to be (*uniquely*) *open-loop solvable at x*. Problem $(SLQ)_\infty$ is said to be (*uniquely*) *open-loop solvable* if it is (uniquely) open-loop solvable at all $x \in \mathbb{R}^n$.

Definition 3.4.2 A pair $(\Theta, v) \in \mathscr{S}[A, C; B, D] \times L_{\mathbb{F}}^2(\mathbb{R}^m)$ is called a *closed-loop strategy* of Problem $(SLQ)_\infty$. The outcome

$$u \triangleq \Theta X + v$$

of a closed-loop strategy (Θ, v) is called a *closed-loop control* for the initial state x, where X is the closed-loop state process corresponding to (x, Θ, v):

$$\begin{cases} dX(t) = \big[(A + B\Theta)X(t) + Bv(t) + b(t)\big]dt \\ \qquad\qquad + \big[(C + D\Theta)X(t) + Dv(t) + \sigma(t)\big]dW(t), \quad t \geqslant 0, \\ X(0) = x. \end{cases}$$

Definition 3.4.3 A closed-loop strategy (Θ^*, v^*) is said to be *optimal* if

$$J(x; \Theta^* X^* + v^*) \leqslant J(x; \Theta X + v),$$

for all $(x, \Theta, v) \in \mathbb{R}^n \times \mathscr{S}[A, C; B, D] \times L_{\mathbb{F}}^2(\mathbb{R}^m)$, where X^* and X are the closed-loop state processes corresponding to (x, Θ^*, v^*) and (x, Θ, v), respectively. If a closed-loop optimal strategy (uniquely) exists, Problem $(SLQ)_\infty$ is said to be *(uniquely) closed-loop solvable*.

It is worth pointing out that, in general, the admissible control sets $\mathcal{U}_{ad}(x)$ are different for different x, and an open-loop optimal control depends on the initial state $x \in \mathbb{R}^n$, whereas a closed-loop optimal strategy is required to be independent of x.

Remark 3.4.4 From Proposition 3.3.3 we see that when system $[A, C; B, D]$ is L^2-stabilizable, the set $\mathcal{U}_{ad}(x)$ of admissible controls is made of closed-loop controls for all x, and similar to Proposition 2.1.5 of Chap. 2, the condition in Definition 3.4.3 is equivalent to the following condition:

$$J(x; \Theta^* X^* + v^*) \leqslant J(x; u), \quad \forall (x, u) \in \mathbb{R}^n \times \mathcal{U}_{ad}(x).$$

As an immediate consequence, the outcome $u^* \equiv \Theta^* X^* + v^*$ of a closed-loop optimal strategy (Θ^*, v^*) is an open-loop optimal control for the initial state $X^*(0)$. Hence, closed-loop solvability implies open-loop solvability.

Recall from Chap. 2 that for LQ optimal control problems in finite horizon, closed-loop solvability implies open-loop solvability, whereas open-loop solvability does not necessarily imply closed-loop solvability. However, for our Problem $(SLQ)_\infty$ (in an infinite horizon), as we shall prove later, the open-loop and closed-loop solvability are equivalent, and both are equivalent to the existence of a *stabilizing solution* to a *generalized algebraic Riccati equation* which will be introduced shortly.

The equivalence between open-loop and closed-loop solvability for Problem $(SLQ)_\infty$ relies on the *stationarity* of Problem $(SLQ)_\infty$. To illustrate, we consider Problem $(SLQ)_\infty^0$, in which the state equation and the cost functional respectively become

$$\begin{cases} dX(t) = [AX(t) + Bu(t)]dt + [CX(t) + Du(t)]dW(t), \quad t \geqslant 0, \\ X(0) = x, \end{cases}$$

$$J^0(x; u) = \mathbb{E} \int_0^\infty \left\langle \begin{pmatrix} Q & S^\top \\ S & R \end{pmatrix} \begin{pmatrix} X(t) \\ u(t) \end{pmatrix}, \begin{pmatrix} X(t) \\ u(t) \end{pmatrix} \right\rangle dt.$$

Suppose the problem is uniquely open-loop solvable. The linear-quadratic structure of Problem $(SLQ)^0_\infty$ then implies that there exists an $\mathbb{R}^{m \times n}$-valued process U^* such that for any initial distribution ξ, $U^*\xi$ is the unique open-loop optimal control. Let X^* be the optimal state process corresponding to the initial state x and the open-loop optimal control $u^* \equiv U^*x$. For fixed but arbitrary $s \geqslant 0$, we can also consider minimizing

$$J^0(s, x; u) = \mathbb{E} \int_s^\infty \left\langle \begin{pmatrix} Q & S^\top \\ S & R \end{pmatrix} \begin{pmatrix} X(t) \\ u(t) \end{pmatrix}, \begin{pmatrix} X(t) \\ u(t) \end{pmatrix} \right\rangle dt$$

subject to

$$\begin{cases} dX(t) = [AX(t) + Bu(t)]dt + [CX(t) + Du(t)]dW(t), & t \geqslant s, \\ X(s) = x. \end{cases}$$

This is still an LQ problem over an infinite time horizon but with initial time being s. We denote it by Problem $(SLQ)^{s,0}_\infty$. Since the matrices A, B, C, D, Q, S, and R are all time-invariant and the time horizons are all infinite, Problem $(SLQ)^0_\infty$ and Problem $(SLQ)^{s,0}_\infty$ can be regarded as the same problem. Thus, $U^*X^*(s)$ is the open-loop optimal control of Problem $(SLQ)^{s,0}_\infty$ for the initial distribution $X^*(s)$. On the other hand, the dynamic programming principle suggests that $U^*(\cdot + s)x$ is also an open-loop optimal control of Problem $(SLQ)^{s,0}_\infty$ for the initial distribution $X^*(s)$. Therefore, one should have

$$U^*(t + s)x = U^*(t)X^*(s), \quad \forall s, t \geqslant 0.$$

In particular, taking $t = 0$, we have

$$u^*(s) = U^*(s)x = U^*(0)X^*(s), \quad s \geqslant 0,$$

which implies the closed-loop solvability of Problem $(SLQ)^0_\infty$.[1]

Definition 3.4.5 The following constrained nonlinear algebraic equation

$$\begin{cases} \mathcal{Q}(P) - \mathcal{S}(P)^\top \mathcal{R}(P)^\dagger \mathcal{S}(P) = 0, \\ \mathscr{R}(\mathcal{S}(P)) \subseteq \mathscr{R}(\mathcal{R}(P)), \\ \mathcal{R}(P) \geqslant 0, \end{cases} \tag{3.4.1}$$

with the unknown $P \in \mathbb{S}^n$, is called a *generalized algebraic Riccati equation* (ARE, for short). A solution P of (3.4.1) is said to be *stabilizing* if there exists a $\Pi \in \mathbb{R}^{m \times n}$ such that the matrix

---

[1] We will place the above heuristic argument on firm mathematical ground in the subsequent sections.

$$\Theta \triangleq -\mathcal{R}(P)^\dagger \mathcal{S}(P) + [I - \mathcal{R}(P)^\dagger \mathcal{R}(P)]\Pi \tag{3.4.2}$$

is a stabilizer of $[A, C; B, D]$.

Remark 3.4.6 If P is a solution (not necessarily stabilizing) to the ARE (3.4.1) and Θ is defined by (3.4.2), then by the properties of the Moore-Penrose pseudoinverse (see Proposition A.1.6 and Remark A.1.7 in Appendix), one has

$$\mathcal{R}(P)\Theta = -\mathcal{S}(P), \quad \mathcal{S}(P)^\top \Theta = -\Theta^\top \mathcal{R}(P)\Theta = -\mathcal{S}(P)^\top \mathcal{R}(P)\mathcal{S}(P).$$

3.5   A Study of Problem $(SLQ)^0_\infty$

In this section we mainly focus on Problem $(SLQ)^0_\infty$, in which the nonhomogeneous terms b, σ, q, and ρ are all zero. In order to simplify the discussion, we assume that the system $[A, C]$ is L^2-stable (i.e., $0 \in \mathcal{S}[A, C; B, D]$); later, we will relax this assumption and extend the results to Problem $(SLQ)_\infty$.

Recall from Proposition 3.3.3 that in the case of $0 \in \mathcal{S}[A, C; B, D]$,

$$\mathcal{U}_{ad}(x) = L^2_{\mathbb{F}}(\mathbb{R}^m), \quad \forall x \in \mathbb{R}^n.$$

This allows us to represent $J(x; u)$ as a quadratic functional on the Hilbert space $L^2_{\mathbb{F}}(\mathbb{R}^m)$.

Proposition 3.5.1 *Suppose that the system $[A, C]$ is L^2-stable. Then there exist a bounded self-adjoint linear operator $M_2 : L^2_{\mathbb{F}}(\mathbb{R}^m) \to L^2_{\mathbb{F}}(\mathbb{R}^m)$, a bounded linear operator $M_1 : \mathbb{R}^n \to L^2_{\mathbb{F}}(\mathbb{R}^m)$, a matrix $M_0 \in \mathbb{S}^n$, and $\hat{u} \in L^2_{\mathbb{F}}(\mathbb{R}^m)$, $\hat{x} \in \mathbb{R}^n$, $c \in \mathbb{R}$ such that for any $(x, u) \in \mathbb{R}^n \times L^2_{\mathbb{F}}(\mathbb{R}^m)$,*

$$J(x; u) = \langle M_2 u, u \rangle + 2\langle M_1 x, u \rangle + \langle M_0 x, x \rangle + 2\langle u, \hat{u} \rangle + 2\langle x, \hat{x} \rangle + c.$$

In particular, in the case of Problem $(SLQ)^0_\infty$ (i.e., $b, \sigma, q, \rho = 0$),

$$J^0(x; u) = \langle M_2 u, u \rangle + 2\langle M_1 x, u \rangle + \langle M_0 x, x \rangle.$$

The proof of Proposition 3.5.1 is similar to the finite horizon case. Such a representation of the cost functional has several consequences, which we summarize as follows.

Proposition 3.5.2 *Suppose that the system $[A, C]$ is L^2-stable. Then the following hold:*

(i) *Problem* $(SLQ)_\infty$ *is open-loop solvable at* x *if and only if* $M_2 \geqslant 0$ *(i.e.,* M_2 *is a positive operator) and* $M_1 x + \hat{u} \in \mathcal{R}(M_2)$. *In this case,* u^* *is an open-loop optimal control for the initial state* x *if and only if*

$$M_2 u^* + M_1 x + \hat{u} = 0.$$

(ii) *If Problem* $(SLQ)_\infty$ *is open-loop solvable, then so is Problem* $(SLQ)^0_\infty$.

(iii) *If Problem* $(SLQ)^0_\infty$ *is open-loop solvable, then there exists a* $U^* \in L^2_{\mathbb{F}}(\mathbb{R}^{m \times n})$ *such that for any* $x \in \mathbb{R}^n$, $U^* x$ *is an open-loop optimal control of Problem* $(SLQ)^0_\infty$ *for the initial state* x.

Proof (i) By definition, a process $u^* \in L^2_{\mathbb{F}}(\mathbb{R}^m)$ is an open-loop optimal control for the initial state x if and only if

$$J(x; u^* + \lambda v) - J(x; u^*) \geqslant 0, \quad \forall v \in L^2_{\mathbb{F}}(\mathbb{R}^m), \ \forall \lambda \in \mathbb{R}. \tag{3.5.1}$$

By Proposition 3.5.1,

$$
\begin{aligned}
J(x; u^* + \lambda v) &= \langle M_2(u^* + \lambda v), u^* + \lambda v \rangle + 2\langle M_1 x, u^* + \lambda v \rangle \\
&\quad + \langle M_0 x, x \rangle + 2\langle u^* + \lambda v, \hat{u} \rangle + 2\langle x, \hat{x} \rangle + c \\
&= J(x; u^*) + \lambda^2 \langle M_2 v, v \rangle + 2\lambda \langle M_2 u^* + M_1 x + \hat{u}, v \rangle.
\end{aligned}
$$

Thus, (3.5.1) is equivalent to

$$\lambda^2 \langle M_2 v, v \rangle + 2\lambda \langle M_2 u^* + M_1 x + \hat{u}, v \rangle \geqslant 0, \quad \forall v \in L^2_{\mathbb{F}}(\mathbb{R}^m), \ \forall \lambda \in \mathbb{R},$$

which in turn is equivalent to

$$\langle M_2 v, v \rangle \geqslant 0, \quad \forall v \in L^2_{\mathbb{F}}(\mathbb{R}^m) \quad \text{and} \quad M_2 u^* + M_1 x + \hat{u} = 0.$$

The conclusions follow readily.

(ii) If Problem $(SLQ)_\infty$ is open-loop solvable, then we have by (i): $M_2 \geqslant 0$ and $M_1 x + \hat{u} \in \mathcal{R}(M_2)$ for all $x \in \mathbb{R}^n$. In particular, by taking $x = 0$, we see that $\hat{u} \in \mathcal{R}(M_2)$, and hence $M_1 x \in \mathcal{R}(M_2)$ for all $x \in \mathbb{R}^n$. Using (i) again, we obtain the open-loop solvability of Problem $(SLQ)^0_\infty$.

(iii) Let $e_1, \ldots, e_n$ be the standard basis for $\mathbb{R}^n$, and let u_i^* be an open-loop optimal control of Problem $(SLQ)^0_\infty$ for the initial state e_i. Then $U^* \triangleq (u_1^*, \ldots, u_n^*)$ has the desired properties. $\qquad\square$

3.5.1 A Finite Horizon Approach

Let Φ be the solution of (3.2.1). If the system $[A, C]$ is L^2-stable, the matrix

$$G \triangleq \mathbb{E} \int_0^\infty \Phi(t)^\top Q \Phi(t) dt$$

is well-defined. Thus, for any $T > 0$, we can consider the following LQ problem over the finite time horizon $[0, T]$.

Problem (SLQ)$_T^0$. For given $x \in \mathbb{R}^n$, find a $u^* \in \mathcal{U}[0, T]$ such that the cost functional

$$J_T^0(x; u) \triangleq \mathbb{E} \left\{ \langle G X_T(T), X_T(T) \rangle + \int_0^T \left\langle \begin{pmatrix} Q & S^\top \\ S & R \end{pmatrix} \begin{pmatrix} X_T(t) \\ u(t) \end{pmatrix}, \begin{pmatrix} X_T(t) \\ u(t) \end{pmatrix} \right\rangle dt \right\}$$

is minimized over $\mathcal{U}[0, T]$, subject to the state equation

$$\begin{cases} dX_T(t) = [AX_T(t) + Bu(t)]dt + [CX_T(t) + Du(t)]dW(t), \\ X_T(0) = x \end{cases} \tag{3.5.2}$$

over the time horizon $[0, T]$.

Proposition 3.5.3 *Suppose that the system $[A, C]$ is L^2-stable. Let $V_T^0(x)$ denote the value function of Problem (SLQ)$_T^0$, and let $V^0(x)$ denote the value function of Problem (SLQ)$_\infty^0$. We have the following results:*

(i) For any $x \in \mathbb{R}^n$ and $u \in \mathcal{U}[0, T]$,

$$J_T^0(x; u) = J^0(x; u_e),$$

where $u_e \in L_\mathbb{F}^2(\mathbb{R}^m)$ is the zero-extension of u:

$$u_e(t) = u(t) \ \ if\ t \in [0, T]; \qquad u_e(t) = 0 \ \ if\ t \in (T, \infty).$$

(ii) If there exists a $\delta > 0$ such that

$$\langle M_2 v, v \rangle \geqslant \delta \|v\|^2, \quad \forall v \in L_\mathbb{F}^2(\mathbb{R}^m), \tag{3.5.3}$$

then

$$J_T^0(0; u) \geqslant \delta \mathbb{E} \int_0^T |u(t)|^2 dt, \quad \forall u \in \mathcal{U}[0, T]. \tag{3.5.4}$$

(iii) $\lim_{T \to \infty} V_T^0(x) = V^0(x)$ for all $x \in \mathbb{R}^n$.

Proof (i) Fix $x \in \mathbb{R}^n$ and take an arbitrary $u \in \mathcal{U}[0, T]$. Let X_T be the solution to (3.5.2) and X be the solution to

$$\begin{cases} dX(t) = [AX(t) + Bu_e(t)]dt + [CX(t) + Du_e(t)]dW(t), & t \geqslant 0, \\ X(0) = x. \end{cases}$$

It is not hard to see that

$$X(t) = \begin{cases} X_T(t), & t \in [0, T], \\ \Phi(t)\Phi(T)^{-1}X_T(T), & t \in (T, \infty). \end{cases}$$

Noting that for $t \geqslant T$, $\Phi(t)\Phi(T)^{-1}$ has the same distribution as $\Phi(t - T)$ and is independent of $\mathcal{F}_T$, we have

$$\mathbb{E}\left\langle \left(\mathbb{E}\int_0^\infty \Phi(t)^\top Q\Phi(t)dt \right) X_T(T), X_T(T) \right\rangle$$

$$= \mathbb{E}\left\langle \left(\mathbb{E}\int_T^\infty \left[\Phi(t - T)^{-1} \right]^\top Q\left[\Phi(t - T)^{-1} \right]dt \right) X_T(T), X_T(T) \right\rangle$$

$$= \mathbb{E}\left\langle \left(\mathbb{E}\int_T^\infty \left[\Phi(t)\Phi(T)^{-1} \right]^\top Q\left[\Phi(t)\Phi(T)^{-1} \right]dt \right) X_T(T), X_T(T) \right\rangle$$

$$= \mathbb{E}\int_T^\infty \left\langle Q\Phi(t)\Phi(T)^{-1}X_T(T), \Phi(t)\Phi(T)^{-1}X_T(T) \right\rangle dt$$

$$= \mathbb{E}\int_T^\infty \left\langle QX(t), X(t) \right\rangle dt.$$

It follows that

$$J_T^0(x; u) = \mathbb{E}\left\{ \left\langle \left(\mathbb{E}\int_0^\infty \Phi(t)^\top Q\Phi(t)dt \right) X_T(T), X_T(T) \right\rangle \right.$$

$$\left. + \int_0^T \left\langle \begin{pmatrix} Q & S^\top \\ S & R \end{pmatrix} \begin{pmatrix} X_T(t) \\ u(t) \end{pmatrix}, \begin{pmatrix} X_T(t) \\ u(t) \end{pmatrix} \right\rangle dt \right\}$$

$$= \mathbb{E}\left\{ \int_T^\infty \left\langle QX, X \right\rangle dt + \int_0^T \left\langle \begin{pmatrix} Q & S^\top \\ S & R \end{pmatrix} \begin{pmatrix} X \\ u_e \end{pmatrix}, \begin{pmatrix} X \\ u_e \end{pmatrix} \right\rangle dt \right\}$$

$$= J^0(x; u_e). \tag{3.5.5}$$

(ii) Taking $x = 0$ in (3.5.5), we obtain

$$J_T^0(0; u) = J^0(0; u_e) = \langle M_2 u_e, u_e \rangle$$

$$\geqslant \delta \, \mathbb{E}\int_0^\infty |u_e(t)|^2 dt = \delta \, \mathbb{E}\int_0^T |u(t)|^2 dt,$$

which proves (ii).

To prove (iii), one first observes that (3.5.5) implies

$$V^0(x) \leqslant J^0(x; u_e) = J_T^0(x; u), \quad \forall u \in \mathcal{U}[0, T].$$

Taking infimum over $u \in \mathcal{U}[0, T]$ yields

$$V^0(x) \leqslant V_T^0(x), \quad \forall T > 0. \tag{3.5.6}$$

On the other hand, if $V^0(x) > -\infty$, then for any given $\varepsilon > 0$, one can find a $u^\varepsilon \in L_{\mathbb{F}}^2(\mathbb{R}^m)$ such that

$$\mathbb{E} \int_0^\infty \left\langle \begin{pmatrix} Q & S^\top \\ S & R \end{pmatrix} \begin{pmatrix} X^\varepsilon(t) \\ u^\varepsilon(t) \end{pmatrix}, \begin{pmatrix} X^\varepsilon(t) \\ u^\varepsilon(t) \end{pmatrix} \right\rangle dt = J^0(x; u^\varepsilon) \leqslant V^0(x) + \varepsilon, \tag{3.5.7}$$

where X^ε is the solution of

$$\begin{cases} dX^\varepsilon(t) = [AX^\varepsilon(t) + Bu^\varepsilon(t)]dt + [CX^\varepsilon(t) + Du^\varepsilon(t)]dW(t), & t \geqslant 0, \\ X^\varepsilon(0) = x. \end{cases}$$

Since by Proposition 3.2.4 $X^\varepsilon \in \mathcal{X}[0, \infty)$, we have for large $T > 0$,

$$|\mathbb{E}\langle GX^\varepsilon(T), X^\varepsilon(T)\rangle| + \left| \mathbb{E} \int_T^\infty \left\langle \begin{pmatrix} Q & S^\top \\ S & R \end{pmatrix} \begin{pmatrix} X^\varepsilon(t) \\ u^\varepsilon(t) \end{pmatrix}, \begin{pmatrix} X^\varepsilon(t) \\ u^\varepsilon(t) \end{pmatrix} \right\rangle dt \right| \leqslant \varepsilon.$$

Let u_T^ε be the restriction of u^ε to $[0, T]$. Then

$$J^0(x; u^\varepsilon) = J_T^0(x; u_T^\varepsilon) - \mathbb{E}\langle GX^\varepsilon(T), X^\varepsilon(T)\rangle$$
$$+ \mathbb{E} \int_T^\infty \left\langle \begin{pmatrix} Q & S^\top \\ S & R \end{pmatrix} \begin{pmatrix} X^\varepsilon(t) \\ u^\varepsilon(t) \end{pmatrix}, \begin{pmatrix} X^\varepsilon(t) \\ u^\varepsilon(t) \end{pmatrix} \right\rangle dt$$
$$\geqslant V_T^0(x) - \varepsilon. \tag{3.5.8}$$

Combining (3.5.7) and (3.5.8), we see that for large $T > 0$,

$$V_T^0(x) \leqslant V^0(x) + 2\varepsilon,$$

which, together with (3.5.6), implies that $V_T^0(x) \to V^0(x)$ as $T \to \infty$. A similar argument applies to the case when $V^0(x) = -\infty$. $\qquad\square$

If the system $[A, C]$ is L^2-stable and the operator M_2 is uniformly positive, i.e., (3.5.3) holds, then Proposition 3.5.1(i) implies that Problem $(SLQ)_\infty^0$ is uniquely open-loop solvable with the unique optimal control given by

$$u_x^* = -M_2^{-1}M_1 x.$$

Substituting the optimal control u^*_x into the cost functional yields

$$V^0(x) = \langle (M_0 - M^*_1 M^{-1}_2 M_1)x, x \rangle, \quad x \in \mathbb{R}^n.$$

Notice that $M_0 - M^*_1 M^{-1}_2 M_1$ is a matrix.

Theorem 3.5.4 *Suppose that the system $[A, C]$ is L^2-stable and that (3.5.3) holds for some $\delta > 0$. Then*

(i) the matrix $P \triangleq M_0 - M^*_1 M^{-1}_2 M_1$ solves the ARE

$$\begin{cases} \mathcal{Q}(P) - \mathcal{S}(P)^\top \mathcal{R}(P)^{-1} \mathcal{S}(P) = 0, \\ \mathcal{R}(P) > 0, \end{cases}$$

(ii) the matrix $\Theta \triangleq -\mathcal{R}(P)^{-1} \mathcal{S}(P)$ is a stabilizer of $[A, C; B, D]$, and
*(iii) the unique open-loop optimal control of Problem $(SLQ)^0_\infty$ for the initial state x
       is given by*

$$u^*_x(t) = \Theta X_\Theta(t; x), \quad t \geqslant 0,$$

where $X_\Theta(\cdot; x)$ is the solution to the closed-loop system

$$\begin{cases} dX(t) = (A + B\Theta)X(t)dt + (C + D\Theta)X(t)dW(t), \quad t \geqslant 0, \\ X(0) = x. \end{cases}$$

Proof By Proposition 3.5.3(i), (3.5.4) holds. This allows us to invoke Theorem 2.5.6 of Chap. 2 to conclude that for any $T > 0$, the differential Riccati equation

$$\begin{cases} \dot{P}_T(t) + \mathcal{Q}(P_T(t)) - \mathcal{S}(P_T(t))^\top \mathcal{R}(P_T(t))^{-1} \mathcal{S}(P_T(t)) = 0, \quad t \in [0, T], \\ P_T(T) = G \end{cases}$$

admits a unique solution $P_T \in C([0, T]; \mathbb{S}^n)$ such that

$$\mathcal{R}(P_T(t)) \geqslant \delta I, \quad \forall t \in [0, T]; \qquad V^0_T(x) = \langle P_T(0)x, x \rangle, \quad \forall x \in \mathbb{R}^n.$$

From Proposition 3.5.3(ii), we see that

$$\lim_{T \to \infty} P_T(0) = P, \quad \mathcal{R}(P) > 0.$$

The first assertion follows by Lemma 3.3.4.

   To prove the last two assertions, we fix an $x \in \mathbb{R}^n$ and let (X^*_x, u^*_x) be the corresponding optimal pair of Problem $(SLQ)^0_\infty$. Applying Itô's formula to $t \mapsto \langle PX^*_x(t), X^*_x(t) \rangle$ and noting that $\lim_{t \to \infty} \langle PX^*_x(t), X^*_x(t) \rangle = 0$, we have

$$-\langle Px, x \rangle = \mathbb{E} \int_0^\infty \Big\{ 2\langle P[AX_x^*(t) + Bu_x^*(t)], X_x^*(t) \rangle$$
$$+ \langle P[CX_x^*(t) + Du_x^*(t)], CX_x^*(t) + Du_x^*(t) \rangle \Big\} dt$$
$$= \mathbb{E} \int_0^\infty \Big\{ \langle (PA + A^\top P + C^\top PC)X_x^*(t), X_x^*(t) \rangle$$
$$+ 2\langle (B^\top P + D^\top PC)X_x^*(t), u_x^*(t) \rangle + \langle D^\top PDu_x^*(t), u_x^*(t) \rangle \Big\} dt.$$

On the other hand, we have

$$\langle Px, x \rangle = J^0(x; u_x^*) = \mathbb{E} \int_0^\infty \Big[\langle QX_x^*, X_x^* \rangle + 2\langle SX_x^*, u_x^* \rangle + \langle Ru_x^*, u_x^* \rangle \Big] dt.$$

Adding the last two equations yields

$$0 = \mathbb{E} \int_0^\infty \Big[\langle \mathcal{Q}(P)X_x^*, X_x^* \rangle + 2\langle \mathcal{S}(P)X_x^*, u_x^* \rangle + \langle \mathcal{R}(P)u_x^*, u_x^* \rangle \Big] dt$$
$$= \mathbb{E} \int_0^\infty \Big[\langle \mathcal{S}(P)^\top \mathcal{R}(P)^{-1}\mathcal{S}(P)X_x^*, X_x^* \rangle + 2\langle \mathcal{S}(P)X_x^*, u_x^* \rangle + \langle \mathcal{R}(P)u_x^*, u_x^* \rangle \Big] dt$$
$$= \mathbb{E} \int_0^\infty \langle \mathcal{R}(P)[u_x^*(t) - \Theta X_x^*(t)], u_x^*(t) - \Theta X_x^*(t) \rangle dt.$$

Since $\mathcal{R}(P) = R + D^\top PD > 0$, we must have

$$u_x^*(t) = \Theta X_x^*(t), \quad t \geqslant 0,$$

and hence X_x^* satisfies

$$\begin{cases} dX_x^*(t) = (A + B\Theta)X_x^*(t)dt + (C + D\Theta)X_x^*(t)dW(t), & t \geqslant 0, \\ X_x^*(0) = x. \end{cases}$$

Since $X_x^* \in \mathcal{X}[0, \infty)$ and x is arbitrary, we conclude that Θ is a stabilizer of $[A, C; B, D]$. The rest of the proof is clear. $\square$

3.5.2 Open-Loop and Closed-Loop Solvability

According to Proposition 3.5.2 and Theorem 3.5.4, the condition $M_2 \geqslant 0$ is merely necessary for the existence of an open-loop optimal control, whereas the uniform positivity condition (3.5.3) is only sufficient. To bridge this gap, consider, for $\varepsilon > 0$, the cost functional

$$J^0_\varepsilon(x; u) \triangleq \mathbb{E} \int_0^\infty \left\langle \begin{pmatrix} Q & S^\top \\ S & R + \varepsilon I \end{pmatrix} \begin{pmatrix} X(t) \\ u(t) \end{pmatrix}, \begin{pmatrix} X(t) \\ u(t) \end{pmatrix} \right\rangle dt$$

$$= J^0(x; u) + \varepsilon \mathbb{E} \int_0^\infty |u(t)|^2 dt$$

$$= \langle (M_2 + \varepsilon I)u, u \rangle + 2\langle M_1 x, u \rangle + \langle M_0 x, x \rangle.$$

Let us denote by Problem $(SLQ)^{0,\varepsilon}_\infty$ the problem of minimizing $J^0_\varepsilon(x; u)$ subject to the state equation

$$\begin{cases} dX(t) = [AX(t) + Bu(t)]dt + [CX(t) + Du(t)]dW(t), & t \geqslant 0, \\ X(0) = x, \end{cases}$$

and by $V^0_\varepsilon(x)$ the corresponding value function. Suppose $M_2 \geqslant 0$. Then the operator $M_2 + \varepsilon I$ is uniformly positive for all $\varepsilon > 0$, and hence Theorem 3.5.4 can be applied to Problem $(SLQ)^{0,\varepsilon}_\infty$. We expect to obtain a characterization for the value function $V^0(x)$ of Problem $(SLQ)^0_\infty$ by letting $\varepsilon \to 0$.

Theorem 3.5.5 *Suppose that the system $[A, C]$ is L^2-stable. If Problem $(SLQ)^0_\infty$ is open-loop solvable, then the generalized ARE (3.4.1) admits a stabilizing solution $P \in \mathbb{S}^n$. Moreover, $V^0(x) = \langle Px, x \rangle$ for all $x \in \mathbb{R}^n$.*

Proof By Proposition 3.5.2, the open-loop solvability of Problem $(SLQ)^0_\infty$ implies $M_2 \geqslant 0$. It follows by Theorem 3.5.4 that for any $\varepsilon > 0$, the ARE

$$\begin{cases} \mathcal{Q}(P_\varepsilon) - \mathcal{S}(P_\varepsilon)^\top [\mathcal{R}(P_\varepsilon) + \varepsilon I]^{-1} \mathcal{S}(P_\varepsilon) = 0, \\ \mathcal{R}(P_\varepsilon) + \varepsilon I > 0 \end{cases} \tag{3.5.9}$$

admits a unique solution $P_\varepsilon \in \mathbb{S}^n$ such that $V^0_\varepsilon(x) = \langle P_\varepsilon x, x \rangle$ for all $x \in \mathbb{R}^n$. Denote

$$\Theta_\varepsilon \triangleq -[\mathcal{R}(P_\varepsilon) + \varepsilon I]^{-1} \mathcal{S}(P_\varepsilon), \tag{3.5.10}$$

which is a stabilizer of $[A, C; B, D]$, and let Ψ_ε be the solution to the matrix SDE

$$\begin{cases} d\Psi_\varepsilon(t) = (A + B\Theta_\varepsilon)\Psi_\varepsilon(t)dt + (C + D\Theta_\varepsilon)\Psi_\varepsilon(t)dW(t), & t \geqslant 0, \\ \Psi_\varepsilon(0) = I. \end{cases}$$

Then the unique open-loop optimal control $u^*_\varepsilon(\cdot; x)$ of Problem $(SLQ)^{0,\varepsilon}_\infty$ for the initial state x is given by

$$u^*_\varepsilon(t; x) = \Theta_\varepsilon \Psi_\varepsilon(t)x, \quad t \geqslant 0.$$

Now let $U^* \in L^2_\mathbb{F}(\mathbb{R}^{m \times n})$ be a process with the property stated in Proposition 3.5.2(iii). By the definition of value function, we have for any $x \in \mathbb{R}^n$ and $\varepsilon > 0$,

$$V^0(x) + \varepsilon\mathbb{E}\int_0^\infty |\Theta_\varepsilon\Psi_\varepsilon(t)x|^2 dt \leqslant J^0(x; \Theta_\varepsilon\Psi_\varepsilon x) + \varepsilon\mathbb{E}\int_0^\infty |\Theta_\varepsilon\Psi_\varepsilon(t)x|^2 dt$$

$$= J_\varepsilon^0(x; \Theta_\varepsilon\Psi_\varepsilon x) = V_\varepsilon^0(x) = \langle P_\varepsilon x, x\rangle \leqslant J_\varepsilon^0(x; U^*x)$$

$$= V^0(x) + \varepsilon\mathbb{E}\int_0^\infty |U^*(t)x|^2 dt. \tag{3.5.11}$$

Equation (3.5.11) implies that for any $x \in \mathbb{R}^n$ and $\varepsilon > 0$,

$$V^0(x) \leqslant \langle P_\varepsilon x, x\rangle \leqslant V^0(x) + \varepsilon\mathbb{E}\int_0^\infty |U^*(t)x|^2 dt, \tag{3.5.12}$$

$$0 \leqslant \mathbb{E}\int_0^\infty |\Theta_\varepsilon\Psi_\varepsilon(t)x|^2 dt \leqslant \mathbb{E}\int_0^\infty |U^*(t)x|^2 dt. \tag{3.5.13}$$

From (3.5.12) we see that $P \equiv \lim_{\varepsilon\to 0} P_\varepsilon$ exists and $V^0(x) = \langle Px, x\rangle$ for all $x \in \mathbb{R}^n$. From (3.5.13) we see that the family of positive semi-definite matrices

$$\Pi_\varepsilon = \mathbb{E}\int_0^\infty \Psi_\varepsilon(t)^\top\Theta_\varepsilon^\top\Theta_\varepsilon\Psi_\varepsilon(t)dt, \quad \varepsilon > 0$$

is bounded. Since Θ_ε is a stabilizer of $[A, C; B, D]$, the system $[A + B\Theta_\varepsilon, C + D\Theta_\varepsilon]$ is L^2-stable. By Theorem 3.2.3, we have

$$\Pi_\varepsilon(A+B\Theta_\varepsilon)+(A+B\Theta_\varepsilon)^\top\Pi_\varepsilon+(C+D\Theta_\varepsilon)^\top\Pi_\varepsilon(C+D\Theta_\varepsilon)+\Theta_\varepsilon^\top\Theta_\varepsilon = 0.$$

It follows that

$$0 \leqslant \Theta_\varepsilon^\top\Theta_\varepsilon \leqslant -[\Pi_\varepsilon(A + B\Theta_\varepsilon) + (A + B\Theta_\varepsilon)^\top\Pi_\varepsilon], \quad \forall\varepsilon > 0.$$

The above, together with the boundedness of $\{\Pi_\varepsilon\}_{\varepsilon>0}$, shows that

$$|\Theta_\varepsilon|^2 \leqslant K(1 + |\Theta_\varepsilon|), \quad \forall\varepsilon > 0, \tag{3.5.14}$$

for some constant $K > 0$. Noting that (3.5.14) implies the boundedness of $\{\Theta_\varepsilon\}_{\varepsilon>0}$, we may choose a sequence $\{\varepsilon_k\}_{k=1}^\infty \subseteq (0, \infty)$ with $\lim_{k\to\infty} \varepsilon_k = 0$ such that $\Theta \equiv \lim_{k\to\infty} \Theta_{\varepsilon_k}$ exists. Observe that

$$\mathcal{R}(P)\Theta = \lim_{k\to\infty}[\mathcal{R}(P_{\varepsilon_k}) + \varepsilon_k I]\Theta_{\varepsilon_k} = -\lim_{k\to\infty} \mathcal{S}(P_{\varepsilon_k}) = -\mathcal{S}(P).$$

Thus, we have by Proposition A.1.6 in Appendix that

$$\mathscr{R}(\mathcal{S}(P)) \subseteq \mathscr{R}(\mathcal{R}(P)), \tag{3.5.15}$$

$$\Theta = -\mathcal{R}(P)^\dagger\mathcal{S}(P) + [I - \mathcal{R}(P)^\dagger\mathcal{R}(P)]\Pi, \tag{3.5.16}$$

for some $\Pi \in \mathbb{R}^{m \times n}$. Notice that by (3.5.10), $\mathcal{S}(P_\varepsilon)^\top = -\Theta_\varepsilon^\top [\mathcal{R}(P_\varepsilon) + \varepsilon I]$. Thus (3.5.9) can be written as

$$\begin{cases} \mathcal{Q}(P_\varepsilon) - \Theta_\varepsilon^\top [\mathcal{R}(P_\varepsilon) + \varepsilon I] \Theta_\varepsilon = 0, \\ \mathcal{R}(P_\varepsilon) + \varepsilon I > 0. \end{cases}$$

Now a passage to the limit along $\{\varepsilon_k\}_{k=1}^\infty$ in the above yields

$$\begin{cases} \mathcal{Q}(P) - \Theta^\top \mathcal{R}(P) \Theta = 0, \\ \mathcal{R}(P) \geqslant 0, \end{cases}$$

which, together with (3.5.15) and (3.5.16), implies that P solves the generalized ARE (3.4.1). To see that P is a stabilizing solution, we need only show $\Theta \in \mathscr{S}[A, C; B, D]$. For this, let Ψ be the solution to the matrix SDE

$$\begin{cases} d\Psi(t) = (A + B\Theta)\Psi(t)dt + (C + D\Theta)\Psi(t)dW(t), \quad t \geqslant 0, \\ \Psi(0) = I. \end{cases}$$

Since $\Theta_{\varepsilon_k} \to \Theta$ as $k \to \infty$, we have $\Psi_{\varepsilon_k}(t) \to \Psi(t)$, a.s. for all $t \geqslant 0$. Using Fatou's lemma and (3.5.13), we obtain

$$\mathbb{E} \int_0^\infty |\Theta\Psi(t)x|^2 dt \leqslant \liminf_{k \to \infty} \mathbb{E} \int_0^\infty |\Theta_{\varepsilon_k}\Psi_{\varepsilon_k}(t)x|^2 dt$$
$$\leqslant \mathbb{E} \int_0^\infty |U^*(t)x|^2 dt < \infty, \quad \forall x \in \mathbb{R}^n.$$

This implies $\Theta\Psi \in L_{\mathbb{F}}^2(\mathbb{R}^{m \times n})$. Thus, by Proposition 3.2.4 $\Psi \in L_{\mathbb{F}}^2(\mathbb{R}^{n \times n})$. Consequently, $\Theta \in \mathscr{S}[A, C; B, D]$. $\square$

When the system $[A, C]$ is L^2-stable, Theorem 3.5.5 shows that the existence of a stabilizing solution to the generalized ARE is necessary for the open-loop solvability of Problem (SLQ)$_\infty^0$. The converse is also true. In fact, we have a stronger result contained in the following proposition.

Proposition 3.5.6 *Suppose that the generalized ARE (3.4.1) admits a stabilizing solution $P \in \mathbb{S}^n$. Then Problem (SLQ)$_\infty^0$ is closed-loop solvable.*

Proof For arbitrary fixed initial state x and admissible control $u \in \mathcal{U}_{ad}(x)$, let $X(\cdot) \equiv X(\cdot\,; x, u)$ be the corresponding solution of the state equation

$$\begin{cases} dX(t) = [AX(t) + Bu(t)]dt + [CX(t) + Du(t)]dW(t), \quad t \geqslant 0, \\ X(0) = x. \end{cases}$$

By applying Itô's formula to $t \mapsto \langle PX(t), X(t) \rangle$, we obtain

$$-\langle Px, x \rangle = \mathbb{E} \int_0^\infty \Big[\langle (PA + A^\top P + C^\top PC)X(t), X(t) \rangle$$
$$+ 2 \langle (B^\top P + D^\top PC)X(t), u(t) \rangle + \langle D^\top PDu(t), u(t) \rangle \Big] dt.$$

It follows that

$$J^0(x; u) - \langle Px, x \rangle = \mathbb{E} \int_0^\infty \left\langle \begin{pmatrix} \mathcal{Q}(P) & \mathcal{S}(P)^\top \\ \mathcal{S}(P) & \mathcal{R}(P) \end{pmatrix} \begin{pmatrix} X(t) \\ u(t) \end{pmatrix}, \begin{pmatrix} X(t) \\ u(t) \end{pmatrix} \right\rangle dt.$$

By the extended Schur's lemma (Appendix, Theorem A.1.8), we have

$$\begin{pmatrix} \mathcal{Q}(P) & \mathcal{S}(P)^\top \\ \mathcal{S}(P) & \mathcal{R}(P) \end{pmatrix} \geqslant 0.$$

Thus,

$$J^0(x; u) \geqslant \langle Px, x \rangle, \quad \forall u \in \mathcal{U}_{ad}(x). \tag{3.5.17}$$

On the other hand, since P is stabilizing, we can choose a $\Pi \in \mathbb{R}^{m \times n}$ such that the matrix

$$\Theta^* \triangleq -\mathcal{R}(P)^\dagger \mathcal{S}(P) + [I - \mathcal{R}(P)^\dagger \mathcal{R}(P)]\Pi$$

is a stabilizer of $[A, C; B, D]$. By Remark 3.4.6,

$$\mathcal{R}(P)\Theta^* = -\mathcal{S}(P), \quad \mathcal{S}(P)^\top \Theta^* = -(\Theta^*)^\top \mathcal{R}(P)\Theta^* = -\mathcal{S}(P)^\top \mathcal{R}(P)\mathcal{S}(P).$$

Thus, for any $x^* \in \mathbb{R}^n$,

$$\left\langle \begin{pmatrix} \mathcal{Q}(P) & \mathcal{S}(P)^\top \\ \mathcal{S}(P) & \mathcal{R}(P) \end{pmatrix} \begin{pmatrix} x^* \\ \Theta^* x^* \end{pmatrix}, \begin{pmatrix} x^* \\ \Theta^* x^* \end{pmatrix} \right\rangle$$
$$= \langle [\mathcal{Q}(P) + 2\mathcal{S}(P)^\top \Theta^* + (\Theta^*)^\top \mathcal{R}(P)\Theta^*]x^*, x^* \rangle$$
$$= \langle [\mathcal{Q}(P) - \mathcal{S}(P)^\top \mathcal{R}(P)^\dagger \mathcal{S}(P)]x^*, x^* \rangle = 0. \tag{3.5.18}$$

We claim that $(\Theta^*, 0)$ is a closed-loop optimal strategy of Problem $(SLQ)_\infty^0$. Indeed, let X^* be the closed-loop state process corresponding to $(x, \Theta^*, 0)$:

$$\begin{cases} dX^*(t) = (A + B\Theta^*)X^*(t)dt + (C + D\Theta^*)X^*(t)dW(t), & t \geqslant 0, \\ X^*(0) = x. \end{cases}$$

Then, applying Itô's rule to $t \mapsto \langle PX^*(t), X^*(t) \rangle$ and using (3.5.18), we have

$$J^0(x; \Theta^* X^*) - \langle Px, x \rangle$$

$$= \mathbb{E} \int_0^\infty \left\langle \begin{pmatrix} \mathcal{Q}(P) & \mathcal{S}(P)^\top \\ \mathcal{S}(P) & \mathcal{R}(P) \end{pmatrix} \begin{pmatrix} X^*(t) \\ \Theta^* X^*(t) \end{pmatrix}, \begin{pmatrix} X^*(t) \\ \Theta^* X^*(t) \end{pmatrix} \right\rangle dt = 0.$$

Since x is arbitrary, the last equation, together with (3.5.17), implies that $(\Theta^*, 0)$ is a closed-loop optimal strategy of Problem (SLQ)$^0_\infty$. $\square$

Remark 3.5.7 From the proof of Proposition 3.5.6, we see that if P is a stabilizing solution to the generalized ARE (3.4.1), then $V^0(x) = \langle Px, x \rangle$ for all $x \in \mathbb{R}^n$. Consequently, the generalized ARE (3.4.1) has at most one stabilizing solution.

Combining Remark 3.4.4, Theorem 3.5.5, and Proposition 3.5.6, we obtain the following result.

Theorem 3.5.8 *Suppose that the system $[A, C]$ is L^2-stable. Then the following statements are equivalent:*

 (i) *Problem (SLQ)$^0_\infty$ is open-loop solvable;*
 (ii) *Problem (SLQ)$^0_\infty$ is closed-loop solvable;*
(iii) *The generalized ARE (3.4.1) admits a unique stabilizing solution.*

3.6 Nonhomogeneous Problems

In this section we return to Problem (SLQ)$_\infty$, in which nonhomogeneous terms appear. We shall prove a result analogous to Theorem 3.5.8 for the general case when **(S)** holds, i.e., $\mathscr{S}[A, C; B, D] \neq \varnothing$. The key idea is to apply Proposition 3.3.3, thus converting Problem (SLQ)$_\infty$ into an equivalent one, in which the corresponding uncontrolled system is L^2-stable.

Let Σ be a stabilizer of $[A, C; B, D]$, and let

$$\begin{cases} \tilde{A} = A + B\Sigma, \quad \tilde{C} = C + D\Sigma, \quad \tilde{S} = S + R\Sigma, \\ \tilde{Q} = Q + S^\top \Sigma + \Sigma^\top S + \Sigma^\top R \Sigma, \quad \tilde{q} = q + \Sigma^\top \rho. \end{cases} \tag{3.6.1}$$

Consider the state equation

$$\begin{cases} d\tilde{X}(t) = \left[\tilde{A} \tilde{X}(t) + Bv(t) + b(t) \right] dt \\ \qquad\qquad + \left[\tilde{C} \tilde{X}(t) + Dv(t) + \sigma(t) \right] dW(t), \quad t \geqslant 0, \\ \tilde{X}(0) = x, \end{cases} \tag{3.6.2}$$

and the cost functional

$$\tilde{J}(x; v) \triangleq J(x; \Sigma \tilde{X} + v)$$

$$= \mathbb{E} \int_0^\infty \left[\left\langle \begin{pmatrix} Q & S^\top \\ S & R \end{pmatrix} \begin{pmatrix} \tilde{X}(t) \\ \Sigma \tilde{X}(t) + v(t) \end{pmatrix}, \begin{pmatrix} \tilde{X}(t) \\ \Sigma \tilde{X}(t) + v(t) \end{pmatrix} \right\rangle \right.$$

$$\left. + 2 \left\langle \begin{pmatrix} q(t) \\ \rho(t) \end{pmatrix}, \begin{pmatrix} \tilde{X}(t) \\ \Sigma \tilde{X}(t) + v(t) \end{pmatrix} \right\rangle \right] dt$$

$$= \mathbb{E} \int_0^\infty \left[\left\langle \begin{pmatrix} \tilde{Q} & \tilde{S}^\top \\ \tilde{S} & R \end{pmatrix} \begin{pmatrix} \tilde{X} \\ v \end{pmatrix}, \begin{pmatrix} \tilde{X} \\ v \end{pmatrix} \right\rangle + 2 \left\langle \begin{pmatrix} \tilde{q} \\ \rho \end{pmatrix}, \begin{pmatrix} \tilde{X} \\ v \end{pmatrix} \right\rangle \right] dt. \qquad (3.6.3)$$

Denote by $\tilde{X}(\cdot\,; x, v)$ the solution of (3.6.2) corresponding to x and v, and by Problem $(\text{SLQ})'_\infty$ the problem of minimizing (3.6.3) subject to (3.6.2). Notice that the system $[\tilde{A}, \tilde{C}]$ is L^2-stable. The following lists several basic facts about Problem $(\text{SLQ})'_\infty$, whose proofs are straightforward consequences of Proposition 3.3.3.

Proposition 3.6.1 *Let Σ be a stabilizer of $[A, C; B, D]$. Then*

(i) *Problem $(\text{SLQ})'_\infty$ is open-loop solvable at $x \in \mathbb{R}^n$ if and only if Problem $(\text{SLQ})_\infty$ is so. In this case, v^* is an open-loop optimal control of Problem $(\text{SLQ})'_\infty$ if and only if $u^* \triangleq v^* + \Sigma \tilde{X}(\cdot\,; x, v^*)$ is an open-loop optimal control of Problem $(\text{SLQ})_\infty$;*

(ii) *Problem $(\text{SLQ})'_\infty$ is closed-loop solvable if and only if Problem $(\text{SLQ})_\infty$ is so. In this case, (Σ^*, v^*) is a closed-loop optimal strategy of Problem $(\text{SLQ})'_\infty$ if and only if $(\Sigma^* + \Sigma, v^*)$ is a closed-loop optimal strategy of Problem $(\text{SLQ})_\infty$.*

We now state the main result of this section.

Theorem 3.6.2 *Let **(S)** hold. Then the following statements are equivalent:*

(i) *Problem $(\text{SLQ})_\infty$ is open-loop solvable;*

(ii) *Problem $(\text{SLQ})_\infty$ is closed-loop solvable;*

(iii) *The generalized ARE (3.4.1) admits a stabilizing solution $P \in \mathbb{S}^n$, and the BSDE*

$$d\eta = -\Big\{ [A - B\mathcal{R}(P)^\dagger \mathcal{S}(P)]^\top \eta + [C - D\mathcal{R}(P)^\dagger \mathcal{S}(P)]^\top \zeta$$

$$+ [C - D\mathcal{R}(P)^\dagger \mathcal{S}(P)]^\top P\sigma - \mathcal{S}(P)^\top \mathcal{R}(P)^\dagger \rho$$

$$+ Pb + q \Big\} dt + \zeta dW, \quad t \geqslant 0, \qquad (3.6.4)$$

admits an L^2-stable adapted solution (η, ζ) such that

$$\theta(t) \triangleq B^\top \eta(t) + D^\top \zeta(t) + D^\top P\sigma(t) + \rho(t) \in \mathscr{R}(\mathcal{R}(P)), \qquad (3.6.5)$$
$$\text{a.e. } t \in [0, \infty), \text{ a.s.}$$

In the above case, all closed-loop optimal strategies (Θ^*, v^*) are given by

$$\begin{cases} \Theta^* = -\mathcal{R}(P)^\dagger \mathcal{S}(P) + [I - \mathcal{R}(P)^\dagger \mathcal{R}(P)]\Pi, \\ v^* = -\mathcal{R}(P)^\dagger \theta + [I - \mathcal{R}(P)^\dagger \mathcal{R}(P)]\nu, \end{cases} \qquad (3.6.6)$$

where $\Pi \in \mathbb{R}^{m \times n}$ is chosen so that $\Theta^* \in \mathscr{S}[A, C; B, D]$ and $v \in L^2_{\mathbb{F}}(\mathbb{R}^m)$ is arbitrary; every open-loop optimal control u^* for the initial state x admits a closed-loop representation:

$$u^*(t) = \Theta^* X^*(t) + v^*(t), \quad t \geqslant 0, \tag{3.6.7}$$

where (Θ^*, v^*) is a closed-loop optimal strategy of Problem $(SLQ)_\infty$ and X^* is the corresponding closed-loop state process. Moreover,

$$
\begin{aligned}
V(x) &= \langle Px, x \rangle + 2\mathbb{E}\langle \eta(0), x \rangle \\
&\quad + \mathbb{E} \int_0^\infty \left[\langle P\sigma, \sigma \rangle + 2\langle \eta, b \rangle + 2\langle \zeta, \sigma \rangle - \langle \mathcal{R}(P)^\dagger \theta, \theta \rangle \right] dt.
\end{aligned}
$$

Before proceeding with the proof, let us make some observations. Suppose that the ARE (3.4.1) admits a stabilizing solution $P \in \mathbb{S}^n$. Then one can choose a matrix $\Pi \in \mathbb{R}^{m \times n}$ such that

$$\Theta \triangleq -\mathcal{R}(P)^\dagger \mathcal{S}(P) + [I - \mathcal{R}(P)^\dagger \mathcal{R}(P)]\Pi \in \mathscr{S}[A, C; B, D].$$

If (η, ζ) is an L^2-stable adapted solution of (3.6.4) satisfying (3.6.5), then there exists a $\vartheta(t)$ such that

$$\theta(t) = \mathcal{R}(P)\vartheta(t), \quad \text{a.e. } t \in [0, \infty), \text{ a.s.}$$

Thus, for a.e. $t \in [0, \infty)$,

$$\left[\Theta^\top + \mathcal{S}(P)^\top \mathcal{R}(P)^\dagger \right]\theta(t) = \Pi^\top \left[I - \mathcal{R}(P)\mathcal{R}(P)^\dagger \right]\mathcal{R}(P)\vartheta(t) = 0, \quad \text{a.s.}$$

It follows that

$$
\begin{aligned}
d\eta &= -\left[A^\top \eta + C^\top \zeta + C^\top P\sigma + Pb + q - \mathcal{S}(P)^\top \mathcal{R}(P)^\dagger \theta \right] dt + \zeta dW \\
&= -\left[A^\top \eta + C^\top \zeta + C^\top P\sigma + Pb + q + \Theta^\top \theta \right] dt + \zeta dW \\
&= -\left[(A+B\Theta)^\top \eta + (C+D\Theta)^\top \zeta + (C+D\Theta)^\top P\sigma + Pb + q + \Theta^\top \rho \right] dt + \zeta dW.
\end{aligned}
$$

Since $[A + B\Theta, C + D\Theta]$ is L^2-stable, we conclude from Theorem A.2.2 in Appendix that the L^2-stable adapted solution of (3.6.4) satisfying (3.6.5) is unique. In particular, when the coefficients b, σ, q, and ρ are all identically zero, $(\eta, \zeta) = (0, 0)$ is the unique solution of (3.6.4) such that (3.6.5) holds. This leads to the following result.

Corollary 3.6.3 *Let* **(S)** *hold. Then the following statements are equivalent:*

(i) Problem $(SLQ)^0_\infty$ is open-loop solvable;
(ii) Problem $(SLQ)^0_\infty$ is closed-loop solvable;
(iii) The generalized ARE (3.4.1) admits a stabilizing solution.

In the above case, the value function of Problem $(SLQ)^0_\infty$ *is given by*

$$V^0(x) = \langle Px, x \rangle, \quad x \in \mathbb{R}^n,$$

and all closed-loop optimal strategies (Θ^*, v^*) *are given by*

$$\Theta^* = -\mathcal{R}(P)^\dagger \mathcal{S}(P) + [I - \mathcal{R}(P)^\dagger \mathcal{R}(P)]\Pi, \quad v^* = [I - \mathcal{R}(P)^\dagger \mathcal{R}(P)]v,$$

where $\Pi \in \mathbb{R}^{m \times n}$ *is chosen so that* $\Theta^* \in \mathscr{S}[A, C; B, D]$ *and* $v \in L^2_{\mathbb{F}}(\mathbb{R}^m)$ *is arbitrary.*

Proof of Theorem 3.6.2. The implication (ii) $\Rightarrow$ (i) follows from Remark 3.4.4.

For the implication (i) $\Rightarrow$ (iii), we begin by considering Problem $(SLQ)'_\infty$, which is open-loop solvable by Proposition 3.6.1(i). Since the system $[\tilde{A}, \tilde{C}]$ is L^2-stable, by Proposition 3.5.2 and Theorem 3.5.5, the ARE

$$\begin{cases} P\tilde{A} + \tilde{A}^\top P + \tilde{C}^\top P\tilde{C} + \tilde{Q} - (PB + \tilde{C}^\top PD + \tilde{S}^\top) \\ \quad \times (R + D^\top PD)^\dagger (B^\top P + D^\top P\tilde{C} + \tilde{S}) = 0, \\ \mathscr{R}(B^\top P + D^\top P\tilde{C} + \tilde{S}) \subseteq \mathscr{R}(R + D^\top PD), \\ \mathcal{R}(P) = R + D^\top PD \geqslant 0 \end{cases} \quad (3.6.8)$$

admits a (unique) stabilizing solution $P \in \mathbb{S}^n$. Choose $\Lambda \in \mathbb{R}^{m \times n}$ such that

$$\Sigma^* \triangleq -\mathcal{R}(P)^\dagger (B^\top P + D^\top P\tilde{C} + \tilde{S}) + [I - \mathcal{R}(P)^\dagger \mathcal{R}(P)]\Lambda$$

is a stabilizer of $[\tilde{A}, \tilde{C}; B, D]$. By Remark 3.4.6 and (3.6.1),

$$\begin{aligned} \mathcal{R}(P)(\Sigma^* + \Sigma) &= -(B^\top P + D^\top P\tilde{C} + \tilde{S}) + \mathcal{R}(P)\Sigma \\ &= -(B^\top P + D^\top PC + S) = \mathcal{S}(P). \end{aligned} \quad (3.6.9)$$

It follows that $\mathscr{R}(\mathcal{S}(P)) \subseteq \mathscr{R}(\mathcal{R}(P))$. Substituting (3.6.1) into the first equation of (3.6.8) gives

$$\begin{aligned} 0 &= \mathcal{Q}(P) + \mathcal{S}(P)^\top \Sigma + \Sigma^\top \mathcal{S}(P) - \mathcal{S}(P)^\top \mathcal{R}(P)^\dagger \mathcal{S}(P) \\ &\quad - \mathcal{S}(P)^\top \mathcal{R}(P)^\dagger \mathcal{R}(P)\Sigma - \Sigma^\top \mathcal{R}(P)\mathcal{R}(P)^\dagger \mathcal{S}(P) \\ &= \mathcal{Q}(P) - \mathcal{S}(P)^\top \mathcal{R}(P)^\dagger \mathcal{S}(P) + \mathcal{S}(P)^\top [I - \mathcal{R}(P)^\dagger \mathcal{R}(P)]\Sigma \\ &\quad + \Sigma^\top [I - \mathcal{R}(P)\mathcal{R}(P)^\dagger]\mathcal{S}(P) \\ &= \mathcal{Q}(P) - \mathcal{S}(P)^\top \mathcal{R}(P)^\dagger \mathcal{S}(P) - (\Sigma^* + \Sigma)^\top \mathcal{R}(P)[I - \mathcal{R}(P)^\dagger \mathcal{R}(P)]\Sigma \\ &\quad - \Sigma^\top [I - \mathcal{R}(P)\mathcal{R}(P)^\dagger]\mathcal{R}(P)(\Sigma^* + \Sigma) \\ &= \mathcal{Q}(P) - \mathcal{S}(P)^\top \mathcal{R}(P)^\dagger \mathcal{S}(P). \end{aligned}$$

Therefore, P solves the ARE (3.4.1). Since $\Sigma^* + \Sigma$ is a stabilizer of $[A, C; B, D]$, we see by (3.6.9) and Proposition A.1.6 in Appendix that P is stabilizing.

Now choose $\Pi \in \mathbb{R}^{m \times n}$ such that the matrix

$$\Theta \triangleq -\mathcal{R}(P)^{\dagger}\mathcal{S}(P) + [I - \mathcal{R}(P)^{\dagger}\mathcal{R}(P)]\Pi$$

is a stabilizer of $[A, C; B, D]$, and consider the following BSDE on $[0, \infty)$:

$$d\eta(t) = -\big[(A + B\Theta)^{\top}\eta + (C + D\Theta)^{\top}\zeta + (C + D\Theta)^{\top}P\sigma$$
$$+ \Theta^{\top}\rho + Pb + q\big]dt + \zeta dW(t). \tag{3.6.10}$$

Since $[A + B\Theta, C + D\Theta]$ is L^2-stable, it follows from Theorem A.2.2 in Appendix that (3.6.10) admits a unique L^2-stable adapted solution (η, ζ). For fixed but arbitrary x and $u \in \mathcal{U}_{ad}(x)$, let $X(\cdot) \equiv X(\cdot\,; x, u)$ be the corresponding state process. Applying Itô's formula to $t \mapsto \langle PX(t), X(t) \rangle$ yields

$$-\langle Px, x \rangle = \mathbb{E}\int_0^{\infty} \Big[\langle (PA + A^{\top}P + C^{\top}PC)X, X \rangle$$
$$+ 2\langle (B^{\top}P + D^{\top}PC)X, u \rangle + \langle D^{\top}PDu, u \rangle$$
$$+ 2\langle C^{\top}P\sigma + Pb, X \rangle + 2\langle D^{\top}P\sigma, u \rangle + \langle P\sigma, \sigma \rangle \Big]dt,$$

and applying Itô's formula to $t \mapsto \langle \eta(t), X(t) \rangle$ yields

$$\mathbb{E}\langle \eta(0), x \rangle = \mathbb{E}\int_0^{\infty} \Big[\langle \Theta^{\top}(B^{\top}\eta + D^{\top}\zeta + D^{\top}P\sigma + \rho), X \rangle$$
$$+ \langle C^{\top}P\sigma + Pb + q, X \rangle - \langle B^{\top}\eta + D^{\top}\zeta, u \rangle - \langle \eta, b \rangle - \langle \zeta, \sigma \rangle \Big]dt.$$

Denote $\theta(t) = B^{\top}\eta(t) + D^{\top}\zeta(t) + D^{\top}P\sigma(t) + \rho(t)$. Then it follows that

$$J(x; u) - \langle Px, x \rangle - 2\mathbb{E}\langle \eta(0), x \rangle$$
$$= \mathbb{E}\int_0^{\infty} \Big[\langle \mathcal{Q}(P)X, X \rangle + 2\langle \mathcal{S}(P)X, u \rangle + \langle \mathcal{R}(P)u, u \rangle$$
$$- 2\langle \Theta^{\top}\theta, X \rangle + 2\langle \theta, u \rangle + \langle P\sigma, \sigma \rangle + 2\langle \eta, b \rangle + 2\langle \zeta, \sigma \rangle \Big]dt$$
$$= \mathbb{E}\int_0^{\infty} \Big[\langle \mathcal{R}(P)(u - \Theta X), u - \Theta X \rangle + 2\langle \theta, u - \Theta X \rangle$$
$$+ \langle P\sigma, \sigma \rangle + 2\langle \eta, b \rangle + 2\langle \zeta, \sigma \rangle \Big]dt. \tag{3.6.11}$$

Let u^* be an open-loop optimal control of Problem $(SLQ)_{\infty}$ for the initial state x, and denote by $X_{\Theta}(\cdot\,; x, v)$ the solution to the following SDE:

$$\begin{cases} dX_\Theta(t) = [(A + B\Theta)X_\Theta(t) + Bv(t) + b(t)]dt \\ \qquad\qquad + [(C + D\Theta)X_\Theta(t) + Dv(t) + \sigma(t)]dW(t), \quad t \geqslant 0, \\ X_\Theta(0) = x. \end{cases}$$

By Proposition 3.3.3, any admissible control with respect to the initial state x is of the form

$$\Theta X_\Theta(\cdot\,; x, v) + v, \quad v \in L^2_\mathbb{F}(\mathbb{R}^m).$$

Thus $u^* = \Theta X_\Theta(\cdot\,; x, v^*) + v^*$ for some $v^* \in L^2_\mathbb{F}(\mathbb{R}^m)$, and

$$\begin{aligned} J(x; \Theta X_\Theta(\cdot\,; x, v^*) + v^*) &= J(x; u^*) \\ &\leqslant J(x; \Theta X_\Theta(\cdot\,; x, v) + v), \quad \forall v \in L^2_\mathbb{F}(\mathbb{R}^m). \end{aligned} \tag{3.6.12}$$

Now taking $u = \Theta X_\Theta(\cdot\,; x, v) + v$ and noting that

$$X(\cdot\,; x, u^*) = X_\Theta(\cdot\,; x, v^*), \quad X(\cdot\,; x, u) = X_\Theta(\cdot\,; x, v),$$

we have from (3.6.11) and (3.6.12) that for any $v \in L^2_\mathbb{F}(\mathbb{R}^m)$,

$$\begin{aligned} \mathbb{E}&\int_0^\infty \big[\langle \mathcal{R}(P)v^*, v^*\rangle + 2\langle \theta, v^*\rangle\big]dt \\ &= J(x; u^*) - \langle Px, x\rangle - 2\mathbb{E}\langle \eta(0), x\rangle - \mathbb{E}\int_0^\infty \big[\langle P\sigma, \sigma\rangle + 2\langle \eta, b\rangle + 2\langle \zeta, \sigma\rangle\big]dt \\ &\leqslant J(x; u) - \langle Px, x\rangle - 2\mathbb{E}\langle \eta(0), x\rangle - \mathbb{E}\int_0^\infty \big[\langle P\sigma, \sigma\rangle + 2\langle \eta, b\rangle + 2\langle \zeta, \sigma\rangle\big]dt \\ &= \mathbb{E}\int_0^\infty \big[\langle \mathcal{R}(P)v, v\rangle + 2\langle \theta, v\rangle\big]dt. \end{aligned}$$

The above inequality implies that v^* is a minimizer of the functional

$$F(v) = \mathbb{E}\int_0^\infty \big[\langle \mathcal{R}(P)v, v\rangle + 2\langle \theta, v\rangle\big]dt, \quad v \in L^2_\mathbb{F}(\mathbb{R}^m).$$

Therefore, we must have

$$\mathcal{R}(P)v^* + \theta = 0, \quad \text{a.e. } t \in [0, \infty), \text{ a.s.}$$

Hence, according to Proposition A.1.6 in Appendix,

$$\begin{cases} \theta \in \mathscr{R}(\mathcal{R}(P)), \text{ and} \\ v^* = -\mathcal{R}(P)^\dagger\theta + [I - \mathcal{R}(P)^\dagger\mathcal{R}(P)]\nu \text{ for some } \nu \in L^2_\mathbb{F}(\mathbb{R}^m). \end{cases}$$

Observing that

$$\left[\Theta^{\mathsf{T}} + \mathcal{S}(P)^{\mathsf{T}}\mathcal{R}(P)^{\dagger}\right]\theta = -\Pi^{\mathsf{T}}\left[I - \mathcal{R}(P)\mathcal{R}(P)^{\dagger}\right]\mathcal{R}(P)v^* = 0,$$

we obtain

$$\begin{aligned}
(A + B\Theta)^{\mathsf{T}}\eta &+ (C + D\Theta)^{\mathsf{T}}\zeta + (C + D\Theta)^{\mathsf{T}}P\sigma + \Theta^{\mathsf{T}}\rho + Pb + q \\
&= A^{\mathsf{T}}\eta + C^{\mathsf{T}}\zeta + C^{\mathsf{T}}P\sigma + Pb + q + \Theta^{\mathsf{T}}\theta \\
&= A^{\mathsf{T}}\eta + C^{\mathsf{T}}\zeta + C^{\mathsf{T}}P\sigma + Pb + q - \mathcal{S}(P)^{\mathsf{T}}\mathcal{R}(P)^{\dagger}\theta \\
&= [A - B\mathcal{R}(P)^{\dagger}\mathcal{S}(P)]^{\mathsf{T}}\eta + [C - D\mathcal{R}(P)^{\dagger}\mathcal{S}(P)]^{\mathsf{T}}\zeta \\
&\quad + [C - D\mathcal{R}(P)^{\dagger}\mathcal{S}(P)]^{\mathsf{T}}P\sigma - \mathcal{S}(P)^{\mathsf{T}}\mathcal{R}(P)^{\dagger}\rho + Pb + q.
\end{aligned}$$

We see then (η, ζ) is an L^2-stable adapted solution of the BSDE (3.6.4). Furthermore, according to Remark A.1.7 in Appendix,

$$\langle\theta, v^*\rangle = -\langle\mathcal{R}(P)v^*, v^*\rangle = -\langle\mathcal{R}(P)^{\dagger}\theta, \theta\rangle.$$

Thus, replacing u by $u^* = \Theta X_{\Theta}(\cdot\,; x, v^*) + v^*$ in (3.6.11) yields

$$\begin{aligned}
V(x) &= J(x; u^*) \\
&= \langle Px, x\rangle + 2\mathbb{E}\langle\eta(0), x\rangle + \mathbb{E}\int_0^{\infty}\left[\langle P\sigma, \sigma\rangle + 2\langle\eta, b\rangle + 2\langle\zeta, \sigma\rangle\right]dt \\
&\quad + \mathbb{E}\int_0^{\infty}\left[\langle\mathcal{R}(P)v^*, v^*\rangle + 2\langle\theta, v^*\rangle\right]dt \\
&= \langle Px, x\rangle + 2\mathbb{E}\langle\eta(0), x\rangle + \mathbb{E}\int_0^{\infty}\left[\langle P\sigma, \sigma\rangle + 2\langle\eta, b\rangle + 2\langle\zeta, \sigma\rangle\right]dt \\
&\quad - \mathbb{E}\int_0^{\infty}\langle\mathcal{R}(P)^{\dagger}\theta, \theta\rangle dt.
\end{aligned}$$

For the implication (iii) $\Rightarrow$ (ii), we take an arbitrary $(x, u) \in \mathbb{R}^n \times \mathcal{U}_{ad}(x)$ and let $X(\cdot) \equiv X(\cdot\,; x, u)$ be the corresponding state process. Proceeding by analogy with (3.6.11), we obtain

$$\begin{aligned}
J(x; u) &= \langle Px, x\rangle + 2\mathbb{E}\langle\eta(0), x\rangle + \mathbb{E}\int_0^{\infty}\left[\langle P\sigma, \sigma\rangle + 2\langle\eta, b\rangle + 2\langle\zeta, \sigma\rangle\right]dt \\
&\quad + \mathbb{E}\int_0^{\infty}\left[\langle\mathcal{Q}(P)X, X\rangle + 2\langle\mathcal{S}(P)X, u\rangle + \langle\mathcal{R}(P)u, u\rangle\right. \\
&\quad\quad \left. + 2\langle\theta, u + \mathcal{R}(P)^{\dagger}\mathcal{S}(P)X\rangle\right]dt. \tag{3.6.13}
\end{aligned}$$

Let (Θ^*, v^*) be defined by (3.6.6). Then by Proposition A.1.6 and Remark A.1.7 in Appendix, we have

$$\mathcal{S}(P) = -\mathcal{R}(P)\Theta^*, \quad \mathcal{Q}(P) = \mathcal{S}(P)\mathcal{R}(P)^\dagger \mathcal{S}(P)^\top = (\Theta^*)^\top \mathcal{R}(P)\Theta^*,$$
$$\theta = -\mathcal{R}(P)v^*, \quad \mathcal{R}(P)\mathcal{R}(P)^\dagger \mathcal{S}(P) = -\mathcal{R}(P)\Theta^*.$$

Substituting the above into (3.6.13) and completing the square, we obtain

$$
\begin{aligned}
J(x; u) = &\langle Px, x \rangle + 2\mathbb{E}\langle \eta(0), x \rangle \\
&+ \mathbb{E}\int_0^\infty \left[\langle P\sigma, \sigma \rangle + 2\langle \eta, b \rangle + 2\langle \zeta, \sigma \rangle - \langle \mathcal{R}(P)v^*, v^* \rangle \right] dt \\
&+ \mathbb{E}\int_0^\infty \langle \mathcal{R}(P)(u - \Theta^* X - v^*), u - \Theta^* X - v^* \rangle dt. \quad (3.6.14)
\end{aligned}
$$

Since $\mathcal{R}(P) \geqslant 0$ and Θ^* is a stabilizer of $[A, C; B, D]$, we have

$$
\begin{aligned}
J(x; u) \geqslant &\langle Px, x \rangle + 2\mathbb{E}\langle \eta(0), x \rangle \\
&+ \mathbb{E}\int_0^\infty \left[\langle P\sigma, \sigma \rangle + 2\langle \eta, b \rangle + 2\langle \zeta, \sigma \rangle - \langle \mathcal{R}(P)v^*, v^* \rangle \right] dt \\
= &J(x; \Theta^* X^* + v^*), \quad \forall x \in \mathbb{R}^n, \ \forall u \in \mathcal{U}_{ad}(x), \quad (3.6.15)
\end{aligned}
$$

which shows (Θ^*, v^*) is a closed-loop optimal strategy of Problem (SLQ)$_\infty$.

Finally, suppose that $(\bar{\Theta}, \bar{v})$ is a closed-loop optimal strategy. Let $\bar{X}$ be the solution to the closed-loop system

$$
\begin{cases}
d\bar{X}(t) = \left[(A + B\bar{\Theta})\bar{X}(t) + B\bar{v}(t) + b(t) \right] dt \\
\qquad\qquad + \left[(C + D\bar{\Theta})\bar{X}(t) + D\bar{v}(t) + \sigma(t) \right] dW(t), \quad t \geqslant 0, \\
\bar{X}(0) = x,
\end{cases}
$$

and let $\bar{u} = \bar{\Theta}\bar{X} + \bar{v}$ denote the outcome of $(\bar{\Theta}, \bar{v})$. Clearly,

$$X(t; x, \bar{u}) = \bar{X}(t), \quad t \geqslant 0.$$

Now (3.6.14) and (3.6.15) imply that

$$
\begin{aligned}
V(x) = J(x; \bar{u}) = &\langle Px, x \rangle + 2\mathbb{E}\langle \eta(0), x \rangle \\
&+ \mathbb{E}\int_0^\infty \left[\langle P\sigma, \sigma \rangle + 2\langle \eta, b \rangle + 2\langle \zeta, \sigma \rangle - \langle \mathcal{R}(P)v^*, v^* \rangle \right] dt \\
&+ \mathbb{E}\int_0^\infty \langle \mathcal{R}(P)(\bar{u} - \Theta^* \bar{X} - v^*), \bar{u} - \Theta^* \bar{X} - v^* \rangle dt \\
= &V(x) + \mathbb{E}\int_0^\infty \left| \mathcal{R}(P)^{\frac{1}{2}} (\bar{\Theta}\bar{X} + \bar{v} - \Theta^* \bar{X} - v^*) \right|^2 dt,
\end{aligned}
$$

from which it follows that

$$\mathcal{R}(P)^{\frac{1}{2}}\big(\bar{\Theta}\bar{X} + \bar{v} - \Theta^*\bar{X} - v^*\big) = 0, \quad \forall x \in \mathbb{R}^n.$$

Multiplying the above by $\mathcal{R}(P)^{\frac{1}{2}}$, we obtain

$$\mathcal{R}(P)(\bar{\Theta} - \Theta^*)\bar{X} + \mathcal{R}(P)(\bar{v} - v^*) = 0, \quad \forall x \in \mathbb{R}^n. \tag{3.6.16}$$

Since (3.6.16) holds for all $x \in \mathbb{R}^n$, and $\bar{\Theta}$, Θ^*, $\bar{v}$, and v^* are independent of x, by subtracting solutions corresponding x and 0, the latter from the former, we see that for any $x \in \mathbb{R}^n$, the solution X_0 of

$$\begin{cases} dX_0(t) = (A + B\bar{\Theta})X_0(t)dt + (C + D\bar{\Theta})X_0(t)dW(t), & t \geqslant 0, \\ X_0(0) = x, \end{cases}$$

satisfies $\mathcal{R}(P)(\bar{\Theta} - \Theta^*)X_0 = 0$, from which we conclude that $\mathcal{R}(P)(\bar{\Theta} - \Theta^*) = 0$ and hence $\mathcal{R}(P)(\bar{v} - v^*) = 0$. Now we have

$$\mathcal{R}(P)\bar{\Theta} = \mathcal{R}(P)\Theta^* = -\mathcal{S}(P), \quad \mathcal{R}(P)\bar{v} = \mathcal{R}(P)v^* = -\theta.$$

According to Proposition A.1.6 in Appendix, $(\bar{\Theta}, \bar{v})$ must be of the form (3.6.6). Similarly, if $\hat{u}$ is an open-loop optimal control for the initial state x, then with $\hat{X}$ denoting the corresponding optimal state process, we have

$$\mathcal{R}(P)(\hat{u} - \Theta^*\hat{X} - v^*) = 0,$$

or equivalently,

$$\mathcal{R}(P)\hat{u} = \mathcal{R}(P)\Theta^*\hat{X} + \mathcal{R}(P)v^* = -\mathcal{S}(P)\hat{X} - \theta.$$

By Proposition A.1.6 in Appendix, there exists a $\nu \in L^2_{\mathbb{F}}(\mathbb{R}^m)$ such that

$$\begin{aligned} \hat{u} &= -\mathcal{R}(P)^\dagger \mathcal{S}(P)\hat{X} - \mathcal{R}(P)^\dagger \theta + [I - \mathcal{R}(P)^\dagger \mathcal{R}(P)]\nu \\ &= \big\{ -\mathcal{R}(P)^\dagger \mathcal{S}(P) + [I - \mathcal{R}(P)^\dagger \mathcal{R}(P)]\Pi \big\}\hat{X} \\ &\quad - \mathcal{R}(P)^\dagger \theta + [I - \mathcal{R}(P)^\dagger \mathcal{R}(P)](\nu - \Pi\hat{X}). \end{aligned}$$

In the above, $\Pi \in \mathbb{R}^{m \times n}$ is chosen such that

$$-\mathcal{R}(P)^\dagger \mathcal{S}(P) + [I - \mathcal{R}(P)^\dagger \mathcal{R}(P)]\Pi \in \mathscr{S}[A, C; B, D].$$

This shows that $\hat{u}$ has the closed-loop representation (3.6.7). $\qquad\square$

3.7   The One-Dimensional Case

In this section we give a complete solution to Problem $(SLQ)^0_\infty$ for the case where both the state and the control variables are one-dimensional, i.e., $n = m = 1$. To avoid trivial exceptions we assume that

$$B^2 + D^2 \neq 0, \quad \mathscr{S}[A, C; B, D] \neq \varnothing. \tag{3.7.1}$$

By Theorem 3.2.3, the second condition in (3.7.1) is equivalent to the solvability of the following inequality for Θ:

$$2(A + B\Theta) + (C + D\Theta)^2 < 0.$$

This inequality admits a solution Θ if and only if

$$(2A + C^2)D^2 < (B + CD)^2. \tag{3.7.2}$$

Let us first look at the case $D = 0$. By scaling, we may assume without loss of generality that $B = 1$. Then the ARE (3.4.1) becomes

$$\begin{cases} P(2A + C^2) + Q - R^\dagger(P + S)^2 = 0, \\ R \geqslant 0, \quad P + S = 0 \text{ if } R = 0. \end{cases} \tag{3.7.3}$$

Also, we note that, by Theorem 3.2.3, Θ is a stabilizer of $[A, C; 1, 0]$ if and only if $\Theta < -(2A + C^2)/2$.

Theorem 3.7.1 *Suppose that $D = 0$ and $B = 1$.*

(i) If $R < 0$, then Problem $(SLQ)^0_\infty$ is not solvable.

(ii) If $R = 0$, then Problem $(SLQ)^0_\infty$ is solvable if and only if $Q = S(2A + C^2)$. In this case, the set of closed-loop optimal strategies is

$$\left\{ (\Theta, v) : \Theta < -(2A + C^2)/2, \, v \in L^2_\mathbb{F}(\mathbb{R}) \right\}.$$

(iii) If $R > 0$, then Problem $(SLQ)^0_\infty$ is solvable if and only if

$$\Sigma \triangleq R(2A + C^2)^2 - 4S(2A + C^2) + 4Q > 0.$$

In this case, $\left(-\dfrac{2A + C^2 + \sqrt{\Sigma/R}}{2}, 0 \right)$ is the unique closed-loop optimal strategy of Problem $(SLQ)^0_\infty$.

Proof (i) It is obvious since $R \geqslant 0$ is violated (see (3.7.3)).

(ii) When $R = 0$, the ARE (3.7.3) further reduces to

$$P(2A + C^2) + Q = 0, \quad P + S = 0,$$

which is solvable if and only if $Q = S(2A + C^2)$. In this case,

$$\mathcal{R}(P) = R = 0, \quad P = -S,$$

and the second assertion follows immediately from Corollary 3.6.3.

(iii) When $R > 0$, the ARE (3.7.3) can be written as

$$P^2 + [2S - (2A + C^2)R]P + S^2 - QR = 0, \qquad (3.7.4)$$

which is solvable if and only if the discriminant

$$\begin{aligned}
\Delta &= [2S - (2A + C^2)R]^2 - 4(S^2 - QR) \\
&= R[R(2A + C^2)^2 - 4S(2A + C^2) + 4Q] \geqslant 0.
\end{aligned}$$

In the case of $\Delta \geqslant 0$, (3.7.4) has two solutions:

$$P_1 = \frac{(2A + C^2)R - 2S - \sqrt{\Delta}}{2}, \quad P_2 = \frac{(2A + C^2)R - 2S + \sqrt{\Delta}}{2},$$

and P_k ($k = 1, 2$) is stabilizing if and only if

$$-\frac{2A + C^2}{2} > -\frac{P_k + S}{R} = -\frac{2A + C^2}{2} - \frac{(-1)^k \sqrt{\Delta}}{2R}.$$

Clearly, P_1 cannot be stabilizing, and P_2 is stabilizing if and only if $\Delta > 0$, or equivalently, $R(2A + C^2)^2 - 4S(2A + C^2) + 4Q > 0$. The second assertion then follows easily.   □

We now look at the case $D \neq 0$. As before, we may assume, without loss of generality (by scaling, if necessary), that $D = 1$. Let

$$\begin{cases}
\alpha = (B + C)^2 - (2A + C^2), \\
\beta = Q - (2A + C^2)R + 2(B + C)[(B + C)R - S], \qquad (3.7.5) \\
\gamma = [(B + C)R - S]^2.
\end{cases}$$

Then (3.7.2) is equivalent to $\alpha > 0$, and Θ is a stabilizer of $[A, C; B, 1]$ if and only if

$$|\Theta + B + C| < \sqrt{\alpha}. \qquad (3.7.6)$$

Theorem 3.7.2 *Suppose that $D = 1$ and $\alpha > 0$. Then Problem (SLQ)$_\infty^0$ is solvable if and only if one of the following conditions holds:*

(i) $Q = (2A + C^2)R$ *and* $S = (B + C)R$. *In this case, the set of closed-loop optimal strategies is*

$$\{(\Theta, v) : |\Theta + B + C| < \sqrt{\alpha}, v \in L_{\mathbb{F}}^2(\mathbb{R})\}. \qquad (3.7.7)$$

(ii) $2A + C^2 \neq 0$, $(2A + C^2)S \geqslant (B + C)Q$, *and*

$$R > \frac{2(B + C - \sqrt{\alpha})S - Q}{(B + C - \sqrt{\alpha})^2}.$$

(iii) $2A + C^2 \neq 0$, $(2A + C^2)S < (B + C)Q$, *and*

$$R > \frac{2(B + C + \sqrt{\alpha})S - Q}{(B + C + \sqrt{\alpha})^2}.$$

(iv) $2A + C^2 = 0$, $Q > 0$, *and*

$$R > \frac{4(B + C)S - Q}{4(B + C)^2}.$$

In the cases (ii), (iii), and (iv),

$$\left(\frac{2\alpha[(B + C)R - S]}{\beta + \sqrt{\beta^2 - 4\alpha\gamma}} - (B + C), \ 0 \right)$$

is the unique closed-loop optimal strategy of Problem $(SLQ)_\infty^0$.

Proof We rewrite the ARE (3.4.1) as follows:

$$\begin{cases} P(2A + C^2) + Q - (R + P)^\dagger [P(B + C) + S]^2 = 0, \\ P(B + C) + S = 0 \quad \text{if } R + P = 0, \\ R + P \geqslant 0. \end{cases} \qquad (3.7.8)$$

By Corollary 3.6.3, Problem $(SLQ)_\infty^0$ is solvable if and only if (3.7.8) admits a stabilizing solution. So we need only discuss the solvability of (3.7.8).

Clearly, $P = -R$ is a solution of (3.7.8) if and only if

$$Q = (2A + C^2)R, \quad S = (B + C)R. \qquad (3.7.9)$$

In this case, $P = -R$ is also stabilizing, and $\mathcal{R}(P) = R + P = 0$. So by Corollary 3.6.3 and (3.7.6), the set of closed-loop optimal strategies of Problem $(SLQ)_\infty^0$ is given by (3.7.7) when (3.7.9) holds.

If (3.7.9) does not hold, by the change of variable $y = R + P$, Eq. (3.7.8) further reduces to

$$\alpha y^2 - \beta y + \gamma = 0, \quad y > 0. \tag{3.7.10}$$

It has a positive solution if and only if

$$\Delta = \beta^2 - 4\alpha\gamma \geqslant 0, \quad \beta + \sqrt{\Delta} > 0,$$

or equivalently (noting that $\alpha > 0$ and $\gamma \geqslant 0$),

$$\Delta = \beta^2 - 4\alpha\gamma \geqslant 0, \quad \beta > 0.$$

In this case, if $\gamma > 0$, then $\Delta < \beta^2$ and hence (3.7.10) has two positive solutions:

$$y_1 = R + P_1 = \frac{\beta - \sqrt{\Delta}}{2\alpha}, \quad y_2 = R + P_2 = \frac{\beta + \sqrt{\Delta}}{2\alpha}. \tag{3.7.11}$$

For $k = 1, 2$, let

$$\Theta_k = -\frac{P_k(B + C) + S}{R + P_k} = \frac{(B + C)R - S}{y_k} - (B + C).$$

Notice that Θ_k is a stabilizer of $[A, C; B, 1]$ if and only if

$$\sqrt{\alpha} > |\Theta_k + B + C| = \left| \frac{(B + C)R - S}{y_k} \right|,$$

which is equivalent to

$$\gamma = [(B + C)R - S]^2 < \alpha y_k^2 = \beta y_k - \gamma. \tag{3.7.12}$$

Upon substitution of (3.7.11) into (3.7.12), the latter in turn is equivalent to

$$\Delta + (-1)^k \beta \sqrt{\Delta} > 0. \tag{3.7.13}$$

Since $\gamma > 0$ (and hence $\Delta < \beta^2$), (3.7.13) cannot hold for $k = 1$, and it holds for $k = 2$ if and only if $\Delta > 0$. Likewise, if $\gamma = 0$, then P_2 is the unique solution of (3.7.10), and Θ_2 is a stabilizer of $[A, C; B, 1]$ if and only if $\Delta > 0$. Therefore, the ARE (3.7.8) admits a stabilizing solution $P \neq R$ if and only if

$$\beta > 0, \quad \beta^2 - 4\alpha\gamma > 0. \tag{3.7.14}$$

Recalling (3.7.5), we have by a straightforward computation:

$$
\begin{aligned}
\beta^2 - 4\alpha\gamma &= [(B+C)^2 - \alpha]^2 R^2 \\
&\quad - \left\{4[(B+C)^2 - \alpha](B+C)S - 2[(B+C)^2 + \alpha]Q\right\}R \\
&\quad + Q^2 - 4(B+C)QS + 4[(B+C)^2 - \alpha]S^2 \\
&\equiv aR^2 - bR + c.
\end{aligned}
$$

Also, we have

$$
b^2 - 4ac = 16\alpha\left\{[(B+C)^2 - \alpha]S - (B+C)Q\right\}^2 \geqslant 0.
$$

If

$$
a = [(B+C)^2 - \alpha]^2 = (2A + C^2)^2 \neq 0,
$$

then

$$
aR^2 - bR + c = \beta^2 - 4\alpha\gamma > 0
$$

if and only if

$$
R > \frac{b + \sqrt{b^2 - 4ac}}{2a} \quad \text{or} \quad R < \frac{b - \sqrt{b^2 - 4ac}}{2a}.
$$

It follows that (3.7.14) holds if and only if one of the following holds:

(1) $R > \dfrac{2(B+C)S - Q}{(B+C)^2 + \alpha}$ and $R > \dfrac{b + \sqrt{b^2 - 4ac}}{2a}$.

(2) $\dfrac{2(B+C)S - Q}{(B+C)^2 + \alpha} < R < \dfrac{b - \sqrt{b^2 - 4ac}}{2a}$.

Because $R = \frac{2(B+C)S-Q}{(B+C)^2+\alpha}$ implies $\beta = 0$ (and hence $aR^2 - bR + c = \beta^2 - 4\alpha\gamma \leqslant 0$), we have

$$
\frac{b - \sqrt{b^2 - 4ac}}{2a} \leqslant \frac{2(B+C)S - Q}{(B+C)^2 + \alpha} \leqslant \frac{b + \sqrt{b^2 - 4ac}}{2a}.
$$

Thus, condition (2) cannot hold. Now, condition (1) is equivalent to

$$
R > \begin{cases} \dfrac{2(B+C-\sqrt{\alpha})S - Q}{(B+C-\sqrt{\alpha})^2}, & \text{if } (2A+C^2)S \geqslant (B+C)Q, \\[3mm] \dfrac{2(B+C+\sqrt{\alpha})S - Q}{(B+C+\sqrt{\alpha})^2}, & \text{if } (2A+C^2)S < (B+C)Q. \end{cases}
$$

If $a = [(B + C)^2 - \alpha]^2 = (2A + C^2)^2 = 0$, then

$$\beta = 2(B + C)^2 R - 2(B + C)S + Q,$$
$$\beta^2 - 4\alpha\gamma = Q[4(B + C)^2 R - 4(B + C)S + Q],$$

and it is not hard to see that (3.7.14) holds if and only if

$$Q > 0, \quad R > \frac{4(B + C)S - Q}{4(B + C)^2}.$$

Finally, in the cases (ii), (iii), and (iv), we see from the preceding argument that the ARE (3.7.8) has a unique stabilizing solution

$$P = \frac{\beta + \sqrt{\Delta}}{2\alpha} - R.$$

Note that $\mathcal{R}(P) = R + P > 0$ and

$$-\mathcal{R}(P)^{-1}\mathcal{S}(P) = -\frac{P(B + C) + S}{R + P} = \frac{2\alpha[(B + C)R - S]}{\beta + \sqrt{\beta^2 - 4\alpha\gamma}} - (B + C).$$

The last assertion follows immediately from Corollary 3.6.3. $\qquad\square$

Appendix
Linear Algebra and BSDEs

A.1 The Moore-Penrose Pseudoinverse

Theorem A.1.1 *For any $M \in \mathbb{R}^{m \times n}$, there exists a unique matrix $M^\dagger \in \mathbb{R}^{n \times m}$ such that*

$$MM^\dagger M = M, \quad (MM^\dagger)^\top = MM^\dagger,$$
$$M^\dagger MM^\dagger = M^\dagger, \quad (M^\dagger M)^\top = M^\dagger M.$$

In addition, if $M \in \mathbb{S}^n$, then $M^\dagger \in \mathbb{S}^n$, $MM^\dagger = M^\dagger M$, and $M \geqslant 0$ if and only if $M^\dagger \geqslant 0$.

The matrix $M^\dagger$ above is called the *Moore-Penrose pseudoinverse* of M.

Example A.1.2 Let $M \in \mathbb{R}^{m \times n}$. Then $(M^\dagger)^\dagger = M$.

Example A.1.3 Let $M = \text{diag}\,(\alpha_1, \ldots, \alpha_m)$ be a diagonal matrix. Then its pseudoinverse is given by $M^\dagger = \text{diag}\,(\beta_1, \ldots, \beta_m)$, where

$$\beta_i = \alpha_i^{-1}, \text{ if } \alpha_i \neq 0; \quad \beta_i = 0, \text{ if } \alpha_i = 0.$$

Proposition A.1.4 *For any $M \in \mathbb{R}^{m \times n}$, we have* $\text{tr}\,(MM^\dagger) \leqslant m$.

Proof By Theorem A.1.1, $A = MM^\dagger$ is symmetric. Thus there exists an orthogonal matrix B such that $BAB^\top = \text{diag}\,(\lambda_1, \ldots, \lambda_k, 0, \ldots, 0)$, where $\lambda_1, \ldots, \lambda_k$ are all the nonzero eigenvalues of A. Then it is not hard to verify that

$$A^\dagger = B^\top \text{diag}\,(\lambda_1^{-1}, \ldots, \lambda_k^{-1}, 0, \ldots, 0)\,B.$$

On the other hand, one can also verify that $A^\dagger = A$. Therefore, with C denoting the diagonal matrix $\text{diag}\,(\lambda_1, \ldots, \lambda_k, 0, \ldots, 0)$, we have

J. Sun and J. Yong, *Stochastic Linear-Quadratic Optimal Control Theory:*
Open-Loop and Closed-Loop Solutions, SpringerBriefs in Mathematics,
https://doi.org/10.1007/978-3-030-20922-3_A

$$A^2 = AA^\dagger = B^\top C B B^\top C^\dagger B = B^\top C C^\dagger B = B^\top \text{diag}\,(1, \ldots, 1, 0, \ldots, 0) B.$$

Now we have the following facts:

$$\text{tr}\,(A) = \lambda_1 + \cdots + \lambda_k, \quad \text{tr}\,(A^2) = \lambda_1^2 + \cdots + \lambda_k^2 = k.$$

Using the Cauchy-Schwarz inequality, we obtain

$$[\text{tr}\,(A)]^2 \leqslant k(\lambda_1^2 + \cdots + \lambda_k^2) = k^2 \leqslant m^2.$$

This completes the proof.   □

Proposition A.1.5 *Let $\mathcal{I}$ be an interval. Let $L(t)$ and $N(t)$ be two Lebesgue measurable functions on $\mathcal{I}$, with values in $\mathbb{R}^{n \times k}$ and $\mathbb{R}^{n \times m}$, respectively. Then the equation $N(t)X(t) = L(t)$ has a solution $X(t) \in L^2(\mathcal{I}; \mathbb{R}^{m \times k})$ if and only if*

(i) $\mathscr{R}(L(t)) \subseteq \mathscr{R}(N(t))$, and
(ii) $N(t)^\dagger L(t) \in L^2(\mathcal{I}; \mathbb{R}^{m \times k})$,

in which case the general solution is given by

$$X(t) = N(t)^\dagger L(t) + [I_m - N(t)^\dagger N(t)]Y(t), \tag{A.1.1}$$

where $Y(t) \in L^2(\mathcal{I}; \mathbb{R}^{m \times k})$ is arbitrary.

Proof Suppose that the equation $N(t)X(t) = L(t)$ has a solution $X(t) \in L^2(\mathcal{I}; \mathbb{R}^{m \times k})$. Then (i) is trivially true. Multiplying both sides of the equation by $N(t)^\dagger$ and making use of Proposition A.1.4, we obtain

$$N(t)^\dagger L(t) = N(t)^\dagger N(t)X(t) \in L^2(\mathcal{I}; \mathbb{R}^{m \times k}).$$

This proves (ii). Observe that $X(t)$ can be written as

$$\begin{aligned} X(t) &= N(t)^\dagger N(t)X(t) + [I_m - N(t)^\dagger N(t)]X(t) \\ &= N(t)^\dagger L(t) + [I_m - N(t)^\dagger N(t)]X(t), \end{aligned}$$

which is of the form (A.1.1).

Conversely, if (i) and (ii) hold, then there exists a function $K(t)$ such that $L(t) = N(t)K(t)$, and for any $Y(t) \in L^2(\mathcal{I}; \mathbb{R}^{m \times k})$, the function $X(t)$ defined by (A.1.1) is in $L^2(\mathcal{I}; \mathbb{R}^{m \times k})$. Since

$$\begin{aligned} N(t)X(t) &= N(t)N(t)^\dagger L(t) + N(t)[I_m - N(t)^\dagger N(t)]Y(t) \\ &= N(t)N(t)^\dagger N(t)K(t) = N(t)K(t) = L(t), \end{aligned}$$

we see that $X(t)$ is a desired solution.   □

Using an argument similar to the one given in the above proof, we can prove the following result.

Proposition A.1.6 *Let $L \in \mathbb{R}^{n \times k}$ and $N \in \mathbb{R}^{n \times m}$. The matrix equation $NX = L$ has a solution if and only if $\mathscr{R}(L) \subseteq \mathscr{R}(N)$, in which case the general solution is given by*

$$X = N^\dagger L + (I_m - N^\dagger N)Y,$$

where $Y \in \mathbb{R}^{m \times k}$ is arbitrary.

Remark A.1.7 The following are obvious:

(i) The condition $\mathscr{R}(L) \subseteq \mathscr{R}(N)$ is equivalent to $NN^\dagger L = L$.
(ii) By Theorem A.1.1, if $N \in \mathbb{S}^n$ and $NX = L$, then $X^\top NX = L^\top N^\dagger L$.

Theorem A.1.8 (Extended Schur's lemma) *Let $L \in \mathbb{R}^{n \times m}$, $M \in \mathbb{S}^n$, and $N \in \mathbb{S}^m$. The following conditions are equivalent:*
(i) $M - LN^\dagger L^\top \geqslant 0$, $N \geqslant 0$, and $\mathscr{R}(L^\top) \subseteq \mathscr{R}(N)$;

(ii) $\begin{pmatrix} M & L \\ L^\top & N \end{pmatrix} \geqslant 0.$

Proof Suppose the condition (i) holds. Then for any $x \in \mathbb{R}^n$ and $y \in \mathbb{R}^m$,

$$
\begin{aligned}
\left(x^\top \ y^\top \right) \begin{pmatrix} M & L \\ L^\top & N \end{pmatrix} \begin{pmatrix} x \\ y \end{pmatrix} &= x^\top M x + 2y^\top L^\top x + y^\top N y \\
&= x^\top (M - LN^\dagger L^\top)x + x^\top LN^\dagger L^\top x + 2y^\top L^\top x + y^\top N y \\
&\geqslant x^\top LN^\dagger L^\top x + 2y^\top L^\top x + y^\top N y.
\end{aligned}
$$

Since $\mathscr{R}(L^\top) \subseteq \mathscr{R}(N)$, there exists a $z \in \mathbb{R}^m$ such that $L^\top x = Nz$. Thus,

$$
\begin{aligned}
x^\top LN^\dagger L^\top x + 2y^\top L^\top x + y^\top N y &= z^\top NN^\dagger Nz + 2y^\top Nz + y^\top N y \\
&= z^\top Nz + 2y^\top Nz + y^\top N y = (y+z)^\top N(y+z) \geqslant 0.
\end{aligned}
$$

Since x and y are arbitrary, (ii) follows by combing the above two inequalities.

Conversely, if (ii) holds, then it is trivially seen that $M \geqslant 0$ and $N \geqslant 0$. Fix an arbitrary $x \in \mathbb{R}^n$ and let $y = -N^\dagger L^\top x$. We have

$$
\begin{aligned}
0 \leqslant \left(x^\top \ y^\top \right) \begin{pmatrix} M & L \\ L^\top & N \end{pmatrix} \begin{pmatrix} x \\ y \end{pmatrix} &= x^\top M x + 2y^\top L^\top x + y^\top N y \\
&= x^\top (M - LN^\dagger L^\top)x,
\end{aligned}
$$

which implies $M - LN^\dagger L^\top \geqslant 0$. Since for any matrix A, the orthogonal complement of its range $\mathscr{R}(A)$ is the kernel of $A^\top$, to prove $\mathscr{R}(L^\top) \subseteq \mathscr{R}(N)$ it suffices to show

$$\mathscr{N}(N) \subseteq \mathscr{N}(L).$$

For this, let $y \in \mathbb{R}^m$ be such that $Ny = 0$. Then for any $x \in \mathbb{R}^m$,

$$0 \leqslant \begin{pmatrix} x^\top & y^\top \end{pmatrix} \begin{pmatrix} M & L \\ L^\top & N \end{pmatrix} \begin{pmatrix} x \\ y \end{pmatrix} = x^\top M x + 2 x^\top L y.$$

Since $M \geqslant 0$ and x is arbitrary, we must have $Ly = 0$.   □

A.2   Linear BSDEs in Infinite Horizons

Let $(\Omega, \mathcal{F}, \mathbb{P})$ be a complete probability space on which a standard one-dimensional Brownian motion $W = \{W(t); 0 \leqslant t < \infty\}$ is defined, and let $\mathbb{F} = \{\mathcal{F}_t\}_{t \geqslant 0}$ be the natural filtration of W augmented by all the $\mathbb{P}$-null sets in $\mathcal{F}$. Consider the following BSDE in the infinite horizon $[0, \infty)$:

$$dY(t) = -\big[A^\top Y(t) + C^\top Z(t) + \varphi(t)\big]dt + Z(t)dW(t), \quad t \in [0, \infty), \quad (A.2.1)$$

where $A, C \in \mathbb{R}^{n \times n}$ are given constant matrices, and $\{\varphi(t); 0 \leqslant t < \infty\}$ is a given $\mathbb{F}$-progressively measurable, $\mathbb{R}^n$-valued process. Recall that $L_{\mathbb{F}}^2(\mathbb{R}^n)$ is the space of $\mathbb{F}$-progressively measurable, $\mathbb{R}^n$-valued processes $\{X(t); 0 \leqslant t < \infty\}$ such that $\mathbb{E} \int_0^\infty |X(t)|^2 dt < \infty$ and that $\mathcal{X}[0, \infty)$ is the subspace of $L_{\mathbb{F}}^2(\mathbb{R}^n)$ whose elements are $\mathbb{F}$-adapted and continuous.

Definition A.2.1 An L^2-*stable adapted solution* to the BSDE (A.2.1) is a pair of processes $(Y, Z) \in \mathcal{X}[0, \infty) \times L_{\mathbb{F}}^2(\mathbb{R}^n)$ satisfying the integral version of (A.2.1):

$$Y(t) = Y(0) - \int_0^t \big[A^\top Y(s) + C^\top Z(s) + \varphi(s)\big]ds$$
$$+ \int_0^t Z(s)dW(s), \quad t \geqslant 0, \quad (A.2.2)$$

almost surely.

Note that if (Y, Z) is an L^2-stable adapted solution to (A.2.1), we must have

$$\lim_{t \to \infty} Y(t) = 0, \quad \text{a.s.}$$

since $\mathbb{E} \int_0^\infty |Y(t)|^2 dt < \infty$. On the other hand, (A.2.2) implies that for any $0 \leqslant T < \infty$,

$$Y(t) = Y(T) + \int_t^T \big[A^\top Y(s) + C^\top Z(s) + \varphi(s)\big]ds - \int_t^T Z(s)dW(s)$$

holds almost surely for all $t \in [0, T]$. Letting $T \to \infty$ then yields

$$Y(t) = \int_t^\infty \left[A^\top Y(s) + C^\top Z(s) + \varphi(s) \right] ds - \int_t^\infty Z(s) dW(s),$$

which resembles the linear BSDE on finite horizon.

We now state the basic existence and uniqueness result for the BSDE (A.2.1). Recall the notion of L^2-stability from Definition 3.2.2 in Chap. 3.

Theorem A.2.2 *Suppose that $[A, C]$ is L^2-stable. Then for any $\varphi \in L^2_{\mathbb{F}}(\mathbb{R}^n)$, equation (A.2.1) admits a unique L^2-stable adapted solution (Y, Z).*

In order to prove Theorem A.2.2, we need the following a priori estimates.

Proposition A.2.3 *Suppose that $[A, C]$ is L^2-stable. Then there exists a constant $K > 0$, depending only on A and C, such that if (Y, Z) is an L^2-stable adapted solution to BSDE (A.2.1), then*

$$\mathbb{E}\left[\sup_{0 \leqslant t < \infty} |Y(t)|^2 \right] + \mathbb{E} \int_0^\infty |Z(t)|^2 dt \leqslant K \mathbb{E} \int_0^\infty |\varphi(t)|^2 dt. \tag{A.2.3}$$

Proof Since $[A, C]$ is L^2-stable, by Theorem 3.2.3 in Chap. 3, there exists a positive definite symmetric matrix P such that $PA + A^\top P + C^\top PC < 0$. One can choose $\varepsilon > 0$ such that the matrix $PA + A^\top P + (1 + \varepsilon)C^\top PC$ is still negative definite. Denote by Π the inverse of P and by Λ the positive definite matrix $-[PA + A^\top P + (1 + \varepsilon)C^\top PC]$. Then we have

$$\Pi \Lambda \Pi = -[A\Pi + \Pi A^\top + (1 + \varepsilon)\Pi C^\top PC\Pi]. \tag{A.2.4}$$

Applying Itô's formula to $s \mapsto \langle \Pi Y(s), Y(s) \rangle$ yields

$$\langle \Pi Y(t), Y(t) \rangle - \langle \Pi Y(T), Y(T) \rangle$$
$$= \int_t^T \left[2\langle \Pi Y, A^\top Y + C^\top Z + \varphi \rangle - \langle \Pi Z, Z \rangle \right] ds - 2 \int_t^T \langle \Pi Y, Z \rangle dW. \tag{A.2.5}$$

Using (A.2.4) we have

$$2\langle \Pi Y, A^\top Y + C^\top Z \rangle = \langle (A\Pi + \Pi A^\top)Y, Y \rangle + 2\langle C\Pi Y, Z \rangle$$
$$= -\langle \Pi \Lambda \Pi Y, Y \rangle - (1 + \varepsilon)\langle \Pi C^\top PC\Pi Y, Y \rangle + 2\langle C\Pi Y, Z \rangle.$$

By completing square and noting that $P > 0$, we obtain

$$(1 + \varepsilon)\langle \Pi C^\top P C \Pi Y, Y \rangle + 2\langle C \Pi Y, Z \rangle$$

$$= (1 + \varepsilon)\left\langle P\left(C\Pi Y - \frac{1}{1+\varepsilon}\Pi Z\right), C\Pi Y - \frac{1}{1+\varepsilon}\Pi Z\right\rangle - \frac{1}{1+\varepsilon}\langle \Pi Z, Z \rangle$$

$$\geqslant -\frac{1}{1+\varepsilon}\langle \Pi Z, Z \rangle,$$

from which it follows that

$$2\langle \Pi Y, A^\top Y + C^\top Z \rangle \leqslant -\langle \Lambda \Pi Y, \Pi Y \rangle + \frac{1}{1+\varepsilon}\langle \Pi Z, Z \rangle.$$

Also by the Cauchy-Schwarz inequality,

$$2\langle \Pi Y, \varphi \rangle \leqslant \langle \Lambda \Pi Y, \Pi Y \rangle + \langle \Lambda^{-1}\varphi, \varphi \rangle.$$

Substituting the last two inequalities into (A.2.5) gives

$$\langle \Pi Y(t), Y(t) \rangle - \langle \Pi Y(T), Y(T) \rangle + \frac{\varepsilon}{1+\varepsilon}\int_t^T \langle \Pi Z, Z \rangle ds$$

$$\leqslant \int_t^T \langle \Lambda^{-1}\varphi, \varphi \rangle ds - 2\int_t^T \langle \Pi Y, Z \rangle dW. \qquad (A.2.6)$$

Let $\lambda > 0$ be the largest eigenvalue of Λ^{-1}, and let ρ_1 and ρ_2 be the smallest and the largest eigenvalues of Π, respectively. Since $Y \in \mathcal{X}[0, \infty)$, we must have $\lim_{T\to\infty} \mathbb{E}|Y(T)|^2 = 0$. Taking expectations on both sides of (A.2.6) and then letting $T \to \infty$, we obtain

$$\rho_1\left[\mathbb{E}|Y(t)|^2 + \frac{\varepsilon}{1+\varepsilon}\mathbb{E}\int_t^\infty |Z(s)|^2 ds\right] \leqslant \lambda\mathbb{E}\int_t^\infty |\varphi(s)|^2 ds.$$

It follows that

$$\mathbb{E}|Y(t)|^2 + \mathbb{E}\int_t^\infty |Z(s)|^2 ds \leqslant \frac{(1+\varepsilon)\lambda}{\varepsilon\rho_1}\mathbb{E}\int_t^\infty |\varphi(s)|^2 ds, \quad \forall t \geqslant 0. \qquad (A.2.7)$$

On the other hand, by the Burkholder-Davis-Gundy inequalities,

$$\mathbb{E}\left[\sup_{0\leqslant t\leqslant T}\left|\int_t^T \langle \Pi Y, Z\rangle dW(s)\right|\right] \leqslant 2\mathbb{E}\left[\sup_{0\leqslant t\leqslant T}\left|\int_0^t \langle \Pi Y, Z\rangle dW(s)\right|\right]$$

$$\leqslant 2\alpha\mathbb{E}\left(\int_0^T |\langle \Pi Y, Z\rangle|^2 ds\right)^{\frac{1}{2}} \leqslant 2\alpha\mathbb{E}\left(\int_0^T |\Pi^{\frac{1}{2}}Y|^2|\Pi^{\frac{1}{2}}Z|^2 ds\right)^{\frac{1}{2}}$$

$$\leqslant 2\alpha\mathbb{E}\left[\left(\sup_{0\leqslant t\leqslant T}|\Pi^{\frac{1}{2}}Y(t)|^2\right)^{\frac{1}{2}}\left(\int_0^T |\Pi^{\frac{1}{2}}Z|^2 ds\right)^{\frac{1}{2}}\right]$$

$$\leqslant \frac{1}{2}\mathbb{E}\left(\sup_{0\leqslant t\leqslant T}|\Pi^{\frac{1}{2}}Y(t)|^2\right) + 2\alpha^2\mathbb{E}\int_0^T |\Pi^{\frac{1}{2}}Z|^2 ds,$$

where $\alpha > 0$ is the universal constant in the Burkholder-Davis-Gundy inequalities and $\Pi^{\frac{1}{2}}$ is the square root of Π. We have used in the last step the Cauchy-Schwarz inequality. This inequality, together with (A.2.6), gives

$$\mathbb{E}\left[\sup_{0\leqslant t\leqslant T}|\Pi^{\frac{1}{2}}Y(t)|^2\right] = \mathbb{E}\left[\sup_{0\leqslant t\leqslant T}\langle \Pi Y(t), Y(t)\rangle\right]$$

$$\leqslant \mathbb{E}\langle \Pi Y(T), Y(T)\rangle + \lambda\mathbb{E}\int_0^T |\varphi(s)|^2 ds$$

$$+ 2\mathbb{E}\left[\sup_{0\leqslant t\leqslant T}\left|\int_t^T \langle \Pi Y(s), Z(s)\rangle dW(s)\right|\right]$$

$$\leqslant \mathbb{E}\langle \Pi Y(T), Y(T)\rangle + \lambda\mathbb{E}\int_0^T |\varphi(s)|^2 ds$$

$$+ \frac{1}{2}\mathbb{E}\left[\sup_{0\leqslant t\leqslant T}|\Pi^{\frac{1}{2}}Y(t)|^2\right] + 2\alpha^2\mathbb{E}\int_0^T \left|\Pi^{\frac{1}{2}}Z(s)\right|^2 ds.$$

Consequently,

$$\mathbb{E}\left[\sup_{0\leqslant t\leqslant T}|\Pi^{\frac{1}{2}}Y(t)|^2\right] \leqslant 2\mathbb{E}\langle \Pi Y(T), Y(T)\rangle + 2\lambda\mathbb{E}\int_0^T |\varphi(s)|^2 ds$$

$$+ 4\alpha^2\mathbb{E}\int_0^T \left|\Pi^{\frac{1}{2}}Z(s)\right|^2 ds.$$

Letting $T \to \infty$ and using the estimate (A.2.7), we obtain

$$\rho_1 \mathbb{E}\left[\sup_{0 \leqslant t \leqslant \infty} |Y(t)|^2\right] \leqslant \mathbb{E}\left[\sup_{0 \leqslant t \leqslant \infty} |\Pi^{\frac{1}{2}} Y(t)|^2\right]$$

$$\leqslant 2\lambda \mathbb{E} \int_0^\infty |\varphi(s)|^2 ds + 4\alpha^2 \mathbb{E} \int_0^\infty |\Pi^{\frac{1}{2}} Z(s)|^2 ds$$

$$\leqslant 2\lambda \mathbb{E} \int_0^\infty |\varphi(s)|^2 ds + 4\alpha^2 \rho_2 \mathbb{E} \int_0^\infty |Z(s)|^2 ds$$

$$\leqslant \left[2\lambda + \frac{4\alpha^2 \lambda \rho_2 (1+\varepsilon)}{\rho_1 \varepsilon}\right] \mathbb{E} \int_0^\infty |\varphi(s)|^2 ds.$$

Combining this estimate and (A.2.7) we obtain the desired result.   □

Proposition A.2.4 *Under the hypotheses of Proposition A.2.3, we also have*

$$\mathbb{E} \int_0^\infty |Y(t)|^2 dt \leqslant K \mathbb{E} \int_0^\infty |\varphi(t)|^2 dt. \tag{A.2.8}$$

Proof Let P and Π be as in the proof of Proposition A.2.3. Applying Itô's formula to $s \mapsto \langle \Pi Y(s), Y(s) \rangle$, we obtain

$$\mathbb{E}\langle \Pi Y(t), Y(t) \rangle - \mathbb{E}\langle \Pi Y(0), Y(0) \rangle$$

$$= \mathbb{E} \int_0^t \left[-2\langle \Pi Y, A^\top Y + C^\top Z + \varphi \rangle + \langle \Pi Z, Z \rangle \right] ds$$

$$\geqslant \mathbb{E} \int_0^t \left[-\langle (A\Pi + \Pi A^\top) Y, Y \rangle - 2\langle \Pi Y, C^\top Z + \varphi \rangle \right] ds.$$

Note that for any $s \geqslant 0$,

$$\langle (A\Pi + \Pi A^\top) Y(s), Y(s) \rangle = \langle (PA + A^\top P) \Pi Y(s), \Pi Y(s) \rangle,$$

and that $PA + A^\top P < 0$. Then with $\mu > 0$ denoting the smallest eigenvalue of $-(PA + A^\top P)$, we have by the Cauchy-Schwarz inequality that

$$\mathbb{E}\langle \Pi Y(t), Y(t) \rangle - \mathbb{E}\langle \Pi Y(0), Y(0) \rangle$$

$$\geqslant \mathbb{E} \int_0^t \left[\mu |\Pi Y(s)|^2 - \frac{\mu}{2} |\Pi Y(s)|^2 - \frac{4}{\mu} |C^\top Z(s)|^2 - \frac{4}{\mu} |\varphi(s)|^2 \right] ds.$$

It follows that (recalling that ρ_1 is the smallest eigenvalues of Π)

$$\rho_1 \mathbb{E} \int_0^t |Y(s)|^2 ds \leqslant \mathbb{E} \int_0^t |\Pi Y(s)|^2 ds$$

$$\leqslant \frac{8}{\mu^2} \mathbb{E} \int_0^t \left[|C^\top Z(s)|^2 + |\varphi(s)|^2 \right] ds + \frac{2}{\mu} \mathbb{E}\langle \Pi Y(t), Y(t) \rangle.$$

Letting $t \to \infty$ and using the a priori estimate (A.2.3), we obtain (A.2.8).   □

Proof of Theorem A.2.2. The uniqueness is an immediate consequence of the a priori estimate (A.2.3). We now prove the existence. For $k = 1, 2, \ldots$, we set

$$\varphi_k(t) \triangleq \varphi(t)\mathbf{1}_{[0,k]}(t), \quad t \in [0, \infty).$$

Clearly, $\{\varphi_k\}_{k=1}^{\infty}$ converges to φ in $L^2_{\mathbb{F}}(\mathbb{R}^n)$. Consider now, for each k, the following BSDE:

$$dY_k(t) = -\big[A^{\top}Y_k(t) + C^{\top}Z_k(t) + \varphi_k(t)\big]dt + Z_k(t)dW(t), \quad t \in [0, \infty).$$

The above BSDE has a unique L^2-stable adapted solution (Y_k, Z_k) which can be constructed as follows: On $[0, k]$, (Y_k, Z_k) is defined to be the adapted solution to the BSDE

$$\begin{cases} dY_k(t) = -\big[A^{\top}Y_k(t) + C^{\top}Z_k(t) + \varphi_k(t)\big]dt + Z_k(t)dW(t), \quad t \in [0, k], \\ Y_k(k) = 0, \end{cases}$$

over the finite horizon $[0, k]$; on (k, ∞), (Y_k, Z_k) is identically equal to zero. According to Propositions A.2.3 and A.2.4, there exists a constant $K > 0$ such that for any $k, j \geqslant 1$

$$\mathbb{E}\left[\sup_{0 \leqslant t < \infty} |Y_k(t) - Y_j(t)|^2\right] + \mathbb{E}\int_0^{\infty} |Y_k(t) - Y_j(t)|^2 dt$$

$$+ \mathbb{E}\int_0^{\infty} |Z_k(t) - Z_j(t)|^2 dt \leqslant K\mathbb{E}\int_0^{\infty} |\varphi_k(t) - \varphi_j(t)|^2 dt.$$

Since $\{\varphi_k\}_{k=1}^{\infty}$ is Cauchy in $L^2_{\mathbb{F}}(\mathbb{R}^n)$, the above inequality implies that $\{(Y_k, Z_k)\}_{k=1}^{\infty}$ has a limit (Y, Z) in $\mathcal{X}[0, \infty) \times L^2_{\mathbb{F}}(\mathbb{R}^n)$. It is not hard to verify that (Y, Z) is an L^2-stable adapted solution to the BSDE (A.2.1).   □

References

1. Ait Rami, M., Moore, J.B., Zhou, X.Y.: Indefinite stochastic linear quadratic control and generalized differential Riccati equation. SIAM J. Control Optim. **40**, 1296–1311 (2001)
2. Ait Rami, M., Zhou, X.Y.: Linear matrix inequalities, Riccati equations, and indefinite stochastic linear quadratic controls. IEEE Trans. Automat. Control **45**, 1131–1143 (2000)
3. Albert, A.: Conditions for positive and nonnegative definiteness in terms of pseudo-inverses. SIAM J. Appl. Math. **17**, 434–440 (1969)
4. Anderson, B.D.O., Moore, J.B.: Optimal Control: Linear Quadratic Methods. Prentice Hall, Englewood Cliffs, NJ (1989)
5. Bank, P., Voss, M.: Linear quadratic stochastic control problems with stochastic terminal constraint. SIAM J. Control Optim. **56**, 672–699 (2018)
6. Bellman, R., Glicksberg, I., Gross, O.: Some Aspects of the Mathematical Theory of Control Processes. RAND Corporation, Santa Monica, CA (1958)
7. Bensoussan, A.: Lectures on stochastic control, part I. In Nonlinear Filtering and Stochastic Control. Lecture Notes in Math, vol. 972, pp. 1–39. Springer, Berlin (1982)
8. Bi, X., Sun, J., Xiong, J.: Optimal control for controllable stochastic linear systems. ESAIM: COCV (2020). https://doi.org/10.1051/cocv/2020027
9. Bismut, J.M.: Linear quadratic optimal stochastic control with random coefficients. SIAM J. Control Optim. **14**, 419–444 (1976)
10. Carmona, R.: Lectures on BSDEs, Stochastic Control, and Stochastic Differential Games with Financial Applications. In: SIAM Book Series in Financial Mathematics, vol. 1 (2016)
11. Chen, S., Li, X., Zhou, X.Y.: Stochastic linear quadratic regulators with indefinite control weight costs. SIAM J. Control Optim. **36**, 1685–1702 (1998)
12. Chen, S., Yong, J.: Stochastic linear quadratic optimal control problems with random coefficients. Chin. Ann. Math. **21 B**, 323–338 (2000)
13. Chen, S., Yong, J.: Stochastic linear quadratic optimal control problems. Appl. Math. Optim. **43**, 21–45 (2001)
14. Chen, S., Zhou, X.Y.: Stochastic linear quadratic regulators with indefinite control weight costs, II. SIAM J. Control Optim. **39**, 1065–1081 (2000)
15. Clements, D., Anderson, B.D.O., Moylan, P.J.: Matrix inequality solution to linear-quadratic singular control problems. IEEE Trans. Automat. Control **22**, 55–57 (1977)
16. Damm, T., Mena, H., Stillfjord, T.: Numerical solution of the finite horizon stochastic linear quadratic control problem. Numer. Linear Algebra Appl. **24**, e2091 (2017)
17. Davis, M.H.A.: Linear Estimation and Stochastic Control. Chapman and Hall, London (1977)
18. Du, K.: Solvability conditions for indefinite linear quadratic optimal stochastic control problems and associated stochastic Riccati equations. SIAM J. Control Optim. **53**, 3673–3689 (2015)

© The Author(s), under exclusive license to Springer Nature Switzerland AG 2020

J. Sun and J. Yong, *Stochastic Linear-Quadratic Optimal Control Theory: Open-Loop and Closed-Loop Solutions*, SpringerBriefs in Mathematics, https://doi.org/10.1007/978-3-030-20922-3

19. Folland, G.B.: Real Analysis: Modern Techniques and Their Applications, 2nd edn. Wiley, New York (1999)
20. Hu, Y., Jin, H., Zhou, X.Y.: Time-inconsistent stochastic linear-quadratic control. SIAM J. Control Optim. **50**, 1548–1572 (2012)
21. Hu, Y., Jin, H., Zhou, X.Y.: Time-inconsistent stochastic linear-quadratic control: characterization and uniqueness of equilibrium. SIAM J. Control Optim. **55**, 1261–1279 (2017)
22. Hu, Y., Peng, S.: Solutions of forward-backward stochastic differential equations. Probab. Theory Related Fields **103**, 273–283 (1995)
23. Huang, J., Li, X., Yong, J.: A linear-quadratic optimal control problem for mean-field stochastic differential equations in infinite horizon. Math. Control Relat. Fields **5**, 97–139 (2015)
24. Kalman, R.E.: Contributions to the theory of optimal control. Bol. Soc. Mat. Mexicana **5**, 102–119 (1960)
25. Karatzas, I., Shreve, S.E.: Brownian Motion and Stochastic Calculus, 2nd edn. Springer, New York (1991)
26. Kohlmann, M., Tang, S.: Minimization of risk and linear quadratic optimal control theory. SIAM J. Control Optim. **42**, 1118–1142 (2003)
27. Letov, A.M.: The analytical design of control systems. Automat. Remote Control **22**, 363–372 (1961)
28. Li, N., Wu, Z., Yu, Z.: Indefinite stochastic linear-quadratic optimal control problems with random jumps and related stochastic Riccati equations. Sci. China Math. **61**, 563–576 (2018)
29. Li, X., Sun, J., Xiong, J.: Linear quadratic optimal control problems for mean-field backward stochastic differential equations. Appl. Math. Optim. **80**, 223–250 (2019)
30. Li, X., Sun, J., Yong, J.: Mean-field stochastic linear quadratic optimal control problems: closed-loop solvability. Probab. Uncertain. Quant. Risk **1**, 2 (2016). https://doi.org/10.1186/s41546-016-0002-3
31. Lim, A.E.B., Zhou, X.Y.: Stochastic optimal LQR control with integral quadratic constraints and indefinite control weights. IEEE Trans. Automat. Control **44**, 1359–1369 (1999)
32. Lü, Q., Wang, T., Zhang, X.: Characterization of optimal feedback for stochastic linear quadratic control problems. Probab. Uncertain. Quant. Risk **2**, 11 (2017). https://doi.org/10.1186/s41546-017-0022-7
33. Ma, J., Protter, P., Yong, J.: Solving forward-backward stochastic differential equations explicitly–a four-step scheme. Probab. Theory Related Fields **98**, 339–359 (1994)
34. Ma, J., Yong, J.: Forward-Backward Stochastic Differential Equations and Their Applications. Lecture Notes in Mathematics, vol. 1702. Springer, New York (1999)
35. McAsey, M., Mou, L.: Generalized Riccati equations arising in stochastic games. Linear Algebra Appl. **416**, 710–723 (2006)
36. Mou, L., Yong, J.: Two-person zero-sum linear quadratic stochastic differential games by a Hilbert space method. J. Ind. Manag. Optim. **2**, 95–117 (2006)
37. Pardoux, E., Tang, S.: Forward-backward stochastic differential equations and quasilinear parabolic PDEs. Probab. Theory Related Fields **114**, 123–150 (1999)
38. Peng, S., Shi, Y.: Infinite horizon forward-backward stochastic differential equations. Stoch. Proc. Appl. **85**, 75–92 (2000)
39. Penrose, R.: A generalized inverse of matrices. Proc. Camb. Philos. Soc. **52**, 17–19 (1955)
40. Pham, H.: Linear quadratic optimal control of conditional McKean-Vlasov equation with random coefficients and applications. Probab. Uncertain. Quant. Risk **1**, 7 (2016). https://doi.org/10.1186/s41546-016-0008-x
41. Pham, H., Basei, M.: Linear-quadratic McKean-Vlasov stochastic control problems with random coefficients on finite and infinite dorizon, and applications (2017). arXiv:1711.09390
42. Qian, Z., Zhou, X.Y.: Existence of solutions to a class of indefinite stochastic Riccati equations. SIAM J. Control Optim. **51**, 221–229 (2013)
43. Sun, J.: Mean-field stochastic linear quadratic optimal control problems: open-loop solvabilities. ESAIM: COCV **23**, 1099–1127 (2017)
44. Sun, J.: Linear quadratic optimal control problems with fixed terminal states and integral quadratic constraints. Appl. Math. Optim. (2018). https://doi.org/10.1007/s00245-018-9532-7

45. Sun, J., Li, X., Yong, J.: Open-loop and closed-loop solvabilities for stochastic linear quadratic optimal control problems. SIAM J. Control Optim. **54**, 2274–2308 (2016)
46. Sun, J., Xiong, J., Yong, J.: Indefinite stochastic linear-quadratic optimal control problems with random coefficients: closed-loop representation of open-loop optimal controls (2019). arXiv:1809.00261v2
47. Sun, J., Yong, J.: Stochastic linear quadratic optimal control problems in infinite horizon. Appl. Math. Optim. **78**, 145–183 (2018)
48. Tang, S.: General linear quadratic optimal stochastic control problems with random coefficients: linear stochastic Hamilton systems and backward stochastic Riccati equations. SIAM J. Control Optim. **42**, 53–75 (2003)
49. Tang, S.: Dynamic programming for general linear quadratic optimal stochastic control with random coefficients. SIAM J. Control Optim. **53**, 1082–1106 (2015)
50. Wang, H., Sun, J., Yong, J.: Weak closed-loop solvability of stochastic linear-quadratic optimal control problems. Discret. Contin. Dyn. Syst. Ser. A **39**, 2785–2805 (2019)
51. Wei, Q., Yong, J., Yu, Z.: Linear quadratic optimal control problems with operator coefficients: open-loop solutions. ESAIM COCV **25**, 17 (2017)
52. Wonham, W.M.: On a matrix Riccati equation of stochastic control. SIAM J. Control **6**, 681–697 (1968)
53. Wu, H., Zhou, X.Y.: Stochastic frequency characteristic. SIAM J. Control Optim. **40**, 557–576 (2001)
54. Yao, D.D., Zhang, S., Zhou, X.Y.: Stochastic linear-quadratic control via semidefinite programming. SIAM J. Control Optim. **40**, 801–823 (2001)
55. Yong, J.: Finding adapted solutions of forward-backward stochastic differential equations–method of continuation. Probab. Theory Related Fields **107**, 537–572 (1997)
56. Yong, J.: Linear forward-backward stochastic differential equations. Appl. Math. Optim. **39**, 93–119 (1999)
57. Yong, J.: Forward backward stochastic differential equations with mixed initial and terminal conditions. Trans. Amer. Math. Soc. **362**, 1047–1096 (2010)
58. Yong, J.: A deterministic linear quadratic time-inconsistent optimal control problem. Math. Control Relat. Fields **1**, 83–118 (2011)
59. Yong, J.: Linear-quadratic optimal control problems for mean-field stochastic differential equations. SIAM J. Control Optim. **51**, 2809–2838 (2013)
60. Yong, J.: Differential Games–A Concise Introduction. World Scientific Publisher, Singapore (2015)
61. Yong, J.: Linear-quadratic optimal control problems for mean-field stochastic differential equations–time-consistent solutions. Trans. Amer. Math. Soc. **369**, 5467–5523 (2017)
62. Yong, J., Zhou, X.Y.: Stochastic Controls: Hamiltonian Systems and HJB Equations. Springer, New York (1999)
63. You, Y.: Optimal control for linear system with quadratic indefinite criterion on Hilbert spaces. Chin. Ann. Math. Ser. **4 B**, 21–32 (1983)

Index

CPSIA information can be obtained
at www.ICGtesting.com
Printed in the USA
LVHW081435010720
659455LV00005B/971

9 783030 209216